ENVIRONMENTAL REMEDIATION TECHNOLOGIES, REGULATIONS AND SAFETY

RIPARIAN ZONES

PROTECTION, RESTORATION AND ECOLOGICAL BENEFITS

ENVIRONMENTAL REMEDIATION TECHNOLOGIES, REGULATIONS AND SAFETY

Additional books in this series can be found on Nova's website under the Series tab.

Additional E-books in this series can be found on Nova's website under the E-book tab.

ENVIRONMENTAL REMEDIATION TECHNOLOGIES, REGULATIONS AND SAFETY

RIPARIAN ZONES

PROTECTION, RESTORATION AND ECOLOGICAL BENEFITS

AHUVA KEREM
AND
HAGAI HAR-EVEN
EDITORS

Nova Science Publishers, Inc.
New York

Library of Congress Cataloging-in-Publication Data

Library of Congress Cataloging-in-Publication Data

Riparian zones : protection, restoration, and ecological benefits / editors, Ahuva Kerem and Hagai Har-Even.
p. cm.
Includes index.
ISBN 978-1-61942-677-1 (hardcover)
1. Riparian areas. 2. Forest protection. 3. Riparian restoration. 4. Riparian ecology. 5. Riparian areas--Management I. Kerem, Ahuva. II. Har-Even, Hagai.
SD411.R57 2011
577.68--dc23

2011052041

Published by Nova Science Publishers, Inc. † New York

CONTENTS

PREFACE

Riparian zones encompass the stream channel and the adjacent portion of the terrestrial landscape from the high water mark toward the uplands, where vegetation might be influenced by elevated water tables or flooding. In this book, the authors present topical research in the study of the protection, restoration and ecological benefits of riparian zones. Topics discussed in this compilation include fire regimes, sand structure, fuel loads and fire behavior in riparian forests of the Sierra Nevada Mountains, USA; the ecological and socioeconomic stakes for conserving and protecting riparian forests in West Africa; ground water chemistry under riparian zones; and protection of riparian landscapes in Israel. (Imprint: Nova).

Chapter 1 – Fire once played an important role in shaping many Sierran coniferous forests, but reduced fire frequency and extent have altered forest conditions. Productive, mesic riparian forests can accumulate high stem densities and fuel loads, making them susceptible to high-severity fire. Fuels treatments applied to upland forests, however, are often excluded from riparian areas due to concerns about degrading forest habitat and water quality. Objectives of this study were to compare fire frequency and seasonality, stand structure, fuel loads, and potential fire behavior between adjacent riparian and upland forests under current and reconstructed conditions. Dendrochronological fire records, current fuel loads, tree diameters, heights, and height to live crown were measured in 36 pairs of riparian and upland plots. Historic estimates of stand structure and fuel loads were reconstructed using equations derived from fuel accumulation rates, current tree data, and increment cores. Fire behavior variables were modeled using Forest Vegetation Simulator Fire/Fuels Extension. Under a liberal filter, riparian fire return intervals (FRIs) ranged from 8.4 to 42.3 years (mean 16.6), while upland FRIs ranged from 6.1 to 58.0 years (mean 16.9). Riparian and upland

FRIs were significantly different in only 1/4 of the sites sampled. Riparian and upland areas did not burn with different seasonalities, and fire events occurred primarily during the late summer-early fall dormant season. FRIs were shorter in forests with a higher proportion of fire-tolerant pine, sites east of the Sierra crest, lower elevation sites, and riparian zones bordering narrower, more incised streams. Upland areas exhibited a greater degree of fire-climate synchrony than riparian areas. Riparian and upland forests were significantly more fire prone under current than reconstructed conditions, with greater basal area, stand density, snag volume, duff loads, total fuel loads, canopy bulk density, surface and crown flame lengths, probability of torching, predicted mortality, and lower torching and crowning indices. Under current conditions, riparian forests were significantly more fire prone than upland forests, with greater stand density, probability of torching, predicted mortality, and lower quadratic mean diameter, canopy base height, and frequency of fire tolerant species. Reconstruction methods found historic riparian and upland forest conditions were not significantly different. Their reconstruction results suggest that historic FRIs, fuels and forest structure may not have differed significantly between many riparian and upland forests. Under current conditions, however, modeled severity is much greater in riparian forests, suggesting habitat and ecosystem function may be more severely impacted by wildfire than in upland forests. Their results emphasize the need for managers to prioritize fuels reduction treatments in Sierra Nevada riparian forests.

Chapter 2 – Riparian forests are vegetation along waterways both in savannah and forest area. In the Dahomey Gap which is the interruption in the West African forest block at the level of Togo and Benin, riparian forests are the main frequently visible forests. In this part of Africa devoid of natural forests, riparian forests play a fundamental role in the conservation of biodiversity, are involved in the ecological functioning of streams, improve water quality, contribute to the bank stabilization of rivers and reduce the impact of sewage and chemical spills from cities. They are also of major socio-economic importance for local populations. These forests along the rivers are still relatively protected by the local populations but intensive agriculture, illegal tree cutting, and charcoal production due to the high demographic growth rate in the region seriously impacted them nowadays. Scientifically, there is an important lack of information specifically relating to the riparian forests of the zone, in which new plant species are regularly assessed. Currently, the potential importance of these riparian forests in the local forestry projects in progress in Togo and Benin, and their patrimonial value for re-afforestation programs, justify the need to set up a monitoring system to get a better knowledge from them. Therefore, this study aim to "document" 1) the physical, social, economic and cultural conditions in which these forests continue to exist, 2) their roles in conservation and protection of habitats and biodiversity, 3) current threats which if left unchecked will lead to their disappearance and, 4) ecological indicators, policies and legal tools to advocate for sustainable management. These information are useful to exploit the potential importance and the phytogenetic value of these riparian forests for monitoring and the local forestry projects.

Chapter 3 – Dairy Arm is a small sand-bed stream whose channel and flood plain were essentially cleared of woody vegetation by early European settlers in the 1820s to 1880s for agriculture. Channel incision of Dairy Arm was initiated on 17-18 June 1949 by a catastrophic flood and resulted in substantial erosion of the channel in the upstream part of the channel network and massive deposition in downstream reaches. River training works carried out in the depositional zone between 1982 and 1991 predominately involved the installation of cylindrical wire mesh barriers for bank protection, stock-proof fencing and the planting of exotic poplars and willows and native river she-oaks. Thirteen permanently marked valley-floor cross sections were surveyed on three occasions between 1981 and 2009. Quantitative data from the cross sections show that the incision cycle initiated by the 1949 flood essentially finished in the early 1990s. Recovery from incision is now well advanced and, in the former incised zone, involves channel contraction, bank revegetation (predominately grasses) and spatially discontinuous bed erosion. Bed erosion involves the creation of a pool-riffle sequence or chain of ponds, and is also very active in the former depositional zone. Channel width has contracted by the construction of a new bench and flood plain in the bed of the former incised channel. In 2009/2010, the vegetation on the bar/bed, bench and flood plain at six cross sections was sampled and the abundance of plant functional types examined. Floristic recovery on Dairy Arm is evidenced by the diversity of plant functional types present on fluvial landforms, by native and exotic species being evenly represented, and by the presence of species indicative of the reference plant community. A lack of catastrophic floods, the exclusion of livestock by river training works, presence of remnant vegetation in the upstream reaches of Dairy Arm (as well as where the channel abuts steep valley slopes), scattered shade trees retained on the flood plain, burial of introduced pasture by substantial sand deposition and the resulting reduction in stocking rates have all played a role in this recovery. Whilst the abundance and diversity of herbaceous plants is good and they have played a crucial role in channel contraction, the same is not the case for woody species. Woody plants are present on the benches, banks and flood plain, but their abundance appears to be limited by grazing and the lack of supply of propagules. The absence of overstorey vegetation is hindering the recovery of important ecological functions, such as the regulation of stream temperature and the supply of large wood for further pool development. Full recovery of both the geomorphic processes and floristic diversity on Dairy Arm will take a long time yet because of the substantial lag time in natural recruitment from planted or remnant vegetation and the production of large wood from overstorey vegetation. This study shows that partial geomorphic and floristic recovery of an incised channel is possible in an agricultural setting, where grazing has been intermittent and stocking densities are low. However, progression to full recovery would be greatly enhanced by protecting the woody vegetation regenerating on the benches, banks and flood plain from livestock.

Chapter 4 – I evaluated ground water chemistry under riparian zones and stream water chemistry in two forested watersheds, northern Japan. The authors selected two small watersheds; 'southern watershed' had an extending riparian wetland, and 'northern watershed' had a narrow riparian area. I measured the stream water chemistry from the upstream (spring) to the downstream (the outlet of the watershed). In longitudinal increase of the stream water flow, stream water chemistry also changed longitudinally in the watersheds. The concentrations of major ions except for ammonium ion (NH_4^+) in both watersheds and chloride ion (Cl^-) in the southern watershed decreased downstream. On the other hand, the concentration of Cl^- and dissolved iron (Fe) in the southern stream increased. NH_4^+ concentrations in the stream water in both watersheds were low (<35 μmolc/L) and did not significantly changed downstream. The downstream decreases of the most solutes were caused by the inflow of the surrounding ground water having low concentrations under the riparian zone to the stream channel. However, Fe concentration in the ground water was higher than in the stream water. No significant change of the Fe concentration in the northern stream suggested the Fe precipitation at the soil-stream interface. Moreover, dissolved organic carbon (DOC) would be related to the nitrogen transformation in the riparian soil and the releases of dissolved Fe, dissolved aluminum (Al) and dissolved silica (Si) from the riparian soil.

Chapter 5 – Riparian landscapes are natural habitats of unique ecological and scenic values, which are highly sensitive to human intervention and impact. Yet, due to their qualities, and especially the presence of water, they are also usually attractive for recreation purposes. This is more so in arid and semi-arid zones like Israel. Nevertheless, in the past, the importance of riparian landscapes in Israel did not receive adequate attention in policy and planning. As a result, over the years they were exposed to various negative impacts, including pollution by industrial and agricultural effluents, exploitation of water for agricultural and other purposes, and land use conflicts. Although in recent years with the growing awareness of their ecological and recreational potential, considerable efforts are being invested in the rehabilitation of deteriorated riparian landscapes, their protection is still deficient. This chapter reviews and examines policy tools used for the protection of riparian landscapes in Israel, based mainly on regulations, reports and existing literature. It concludes by offering some lessons for policy-making in general and suggestions for improving the protection of riparian landscapes in Israel in particular.

Chapter 6 – The endomycorrhizal association established between arbuscular mycorrhizal fungi (Glomeromycetes) and plant roots are known to have a pivotal role on plant survival and nutrition. In this chapter, the authors convey that the triad riparian forest-carbon sequestration-arbuscular mycorrhizal fungi should be considered as environmental friendly strategy to recovery riparian sites. Greenhouse and field experiments carried out in soils occurring at the Itajai valley basin have demonstrated the importance of this association for growth and carbon allocation on plants species used to revegetation of riparian sites. Inoculation of pioneer species with selected fungal isolates have shown that more carbon is allocated to roots when compared to non-inoculated plants or plants inoculated with indigenous fungi. For Schinus terebinthifolius they have evidence at field conditions that inoculation with Scutellospora heterogama, Glomus clarum, and Acaulospora koskei increased significantly the quantity of carbon stored in individual plants, calculated using allometric equations. Riparian sites are habitats that harbors a great diversity of glomeromycetes and 42 species has been registered in works developed in the USA and Brazil, most of them in the genera Glomus and Acaulospora. It is also noticeable that all members of Paraglomus (P. occultum, P. brasilianum, P. laccatum) and Archaeospora trappei are commonly found in riparian sites. Considering the role of arbuscular mycorrhizal fungi on growth and carbon accumulation of pioneer plants and the high diversity of glomeromycetes in riparian soils, the inoculation of plants to be used in revegetation process of riparian sites should be viewed as a environmental friendly strategy to recovery these areas and improve carbon sequestration.

In: Riparian Zones
Editor: A. Kerem et al.
ISBN 978-1-61942-677-1

Chapter 1

FIRE REGIMES, STAND STRUCTURE, FUEL LOADS, AND FIRE BEHAVIOR IN RIPARIAN AND UPLAND FORESTS, SIERRA NEVADA MOUNTAINS, USA

***Kip Van de Water*[1,*] *and Malcolm North*[2,3]**
USDA Forest Service, Plumas National Forest, Feather River Ranger District, 875 Mitchell Avenue, Oroville, CA 95965[1]
USDA Forest Service Pacific Southwest Research Station, 1731 Research Park Dr., Davis, CA 95618[2]
Department of Plant Sciences, One Shields Ave., University of California 95616[3]

ABSTRACT

Fire once played an important role in shaping many Sierran coniferous forests, but reduced fire frequency and extent have altered forest conditions. Productive, mesic riparian forests can accumulate high stem densities and fuel loads, making them susceptible to high-severity fire. Fuels treatments applied to upland forests, however, are often excluded from riparian areas due to concerns about degrading forest habitat and water quality. Objectives of this study were to compare fire frequency and seasonality, stand structure, fuel loads, and potential fire behavior between adjacent riparian and upland

* Corresponding author: telephone 805-455-4065; fax 530-752-1819; email kvandewater@fs.fed.us

forests under current and reconstructed conditions. Dendrochronological fire records, current fuel loads, tree diameters, heights, and height to live crown were measured in 36 pairs of riparian and upland plots. Historic estimates of stand structure and fuel loads were reconstructed using equations derived from fuel accumulation rates, current tree data, and increment cores. Fire behavior variables were modeled using Forest Vegetation Simulator Fire/Fuels Extension.

Under a liberal filter, riparian fire return intervals (FRIs) ranged from 8.4 to 42.3 years (mean 16.6), while upland FRIs ranged from 6.1 to 58.0 years (mean 16.9). Riparian and upland FRIs were significantly different in only 1/4 of the sites sampled. Riparian and upland areas did not burn with different seasonalities, and fire events occurred primarily during the late summer-early fall dormant season. FRIs were shorter in forests with a higher proportion of fire-tolerant pine, sites east of the Sierra crest, lower elevation sites, and riparian zones bordering narrower, more incised streams. Upland areas exhibited a greater degree of fire-climate synchrony than riparian areas.

Riparian and upland forests were significantly more fire prone under current than reconstructed conditions, with greater basal area, stand density, snag volume, duff loads, total fuel loads, canopy bulk density, surface and crown flame lengths, probability of torching, predicted mortality, and lower torching and crowning indices. Under current conditions, riparian forests were significantly more fire prone than upland forests, with greater stand density, probability of torching, predicted mortality, and lower quadratic mean diameter, canopy base height, and frequency of fire tolerant species. Reconstruction methods found historic riparian and upland forest conditions were not significantly different. Our reconstruction results suggest that historic FRIs, fuels and forest structure may not have differed significantly between many riparian and upland forests. Under current conditions, however, modeled severity is much greater in riparian forests, suggesting habitat and ecosystem function may be more severely impacted by wildfire than in upland forests. Our results emphasize the need for managers to prioritize fuels reduction treatments in Sierra Nevada riparian forests.

1. Introduction

Fire plays an important role in shaping stand structure, species composition, and fuel loads in many Sierran coniferous forest types. However, longer fire return intervals and reductions in annual area burned caused by fire suppression and changes in climate and grazing practices have altered forest conditions (Anderson and Moratto, 1996; Douglass and Bilbao, 1975; Dwire and Kauffman, 2003; Pyne, 1982; Skinner and Chang, 1996; Stephens et al., 2007). High densities of small trees and increased fuel loads are now present in many productive forest types that were historically maintained by frequent low- to moderate-intensity fires, resulting in increased risk of high-intensity fire (McKelvey and Busse, 1996; Stephens and Moghaddas, 2005). Development in wildland-urban interface areas with high fuel loads continues at an increasing rate, spurring land managers to suppress most wildfires despite policies that encourage reintroduction of fire as an ecosystem process (Jensen and McPherson, 2008).

Although fuel reduction is accomplished in strategic areas using treatments such as mechanical thinning and prescribed burning, treatment has historically been limited or excluded from riparian areas (FEMAT, 1993; UDSA, 2004; McCaffery et al., 2008; Safford et al., 2009). While active management of fuels in riparian areas is becoming increasingly common (Stone et al., 2010), there is a lack of riparian stand structure and fuel load data that could support the perceived need for riparian fuels management. Riparian forests are often very productive due to greater moisture availability and have accumulated high stem densities and fuel loads, making them susceptible to high-severity fire and subsequent stream channel erosion, loss of wildlife habitat, and decreased ecosystem function (Camp et al., 1997; Olson and Agee, 2005; Segura and Snook, 1992; Skinner and Chang, 1996). High fuel loads and stem densities in riparian forests may allow them to act as a wick for high-intensity fire through a landscape of treated upland forest (e.g. the 2007 Angora Fire in the Tahoe Basin) under some conditions (Murphy et al., 2007; Pettit and Naiman, 2007). Although riparian areas are characterized by lower temperatures and higher humidity (Rambo and North 2008, 2009) than adjacent upland areas, which may slow fire spread through the landscape under non-drought conditions (Skinner and Chang, 1996), they often burn at similar frequencies and may even propagate fire through the upland matrix during extreme weather conditions (Agee, 1998; Dwire and Kauffman, 2003; Pettit and Naiman, 2007). Despite increasing recognition of the importance of fire in some riparian forests, few studies have attempted to reconstruct historic riparian stand structure and fuel loads in the context of an active fire regime (Poage, 1994). Assessing the relationship between current and historic fire regimes, stand structure and fuel loads in adjacent riparian and upland forests could be useful in guiding efforts to restore forest ecosystems altered by fire exclusion and past timber harvesting.

Objectives of this study were to determine whether adjacent coniferous riparian and upland forests historically burned with different frequencies and seasonalities, whether the relationship varied by forest, site and stream characteristics, whether fire synchrony with climate conditions differed between riparian and upland forests, whether current riparian and upland forests have different stand structure, fuel loads, and potential fire behavior than historic riparian and upland forests, and whether riparian forests currently or historically had different stand structure, fuel loads, and potential fire behavior than upland forests. Because few studies of the linkages between fire and riparian stand dynamics have been conducted, additional objectives were to explore the relationships between historic stand conditions and fire regimes, as well as between riparian and upland forests under current versus historic conditions.

We hypothesized that: (1) riparian areas would have longer fire return intervals than adjacent upland forests; (2) riparian areas would have a lower proportion of non-dormant season fires than adjacent upland areas; (3) riparian and upland fire return intervals would be shorter in sites with a greater percentage of the species composition comprised of fire-tolerant pine; (4) riparian and upland fire return intervals would be longer in higher-precipitation (west side) forests than in lower-precipitation (east side) forests; (5) riparian and upland fire return intervals would be shorter in sites with steeper slope, south-facing aspect, and lower elevation; (6) riparian fire return intervals would be longer in sites with a broad channel shape, a perennial flow regime, and greater stream channel width, depth, width to depth ratio, and lower gradient; (7) riparian forests would show greater synchrony between fire events and extreme drought conditions than upland forests; (8) both riparian and upland forests would show a reduction in fire return interval after Euro-American settlement; (9) current riparian and upland stands would have stand structure and fuel loads more conducive to high-intensity fire than reconstructed riparian and upland stands; (10) current riparian stands would have stand structure and fuel loads more conducive to high-intensity fire than current upland stands; and (11) reconstructed riparian stands would have stand structure, fuel loads and potential fire behavior similar to reconstructed upland stands. Attributes suggesting that a stand is conducive to high-intensity fire include high basal area, stand density, snag volume, fuel loads, flame length, probability of torching, canopy bulk density, and potential mortality, and low quadratic mean diameter, canopy base height, fire-tolerant species composition, torching index, and crowning index. Additionally, we investigated the correlation between: (a) fire return intervals and reconstructed stand structure, fuel loads, and predicted fire behavior in riparian and upland forests; and (b) riparian and upland stand structure, fuel loads, and predicted fire behavior under current and reconstructed conditions.

2. METHODS

2.1. Study area and Site Selection

Sampling occurred in four areas of the northern Sierra Nevada: the Almanor Ranger District of the Lassen National Forest (15 sites), the Onion Creek Experimental Forest (4 sites), and the east and west sides of Lake Tahoe Basin (6 and 11 sites, respectively) (Figure 1).

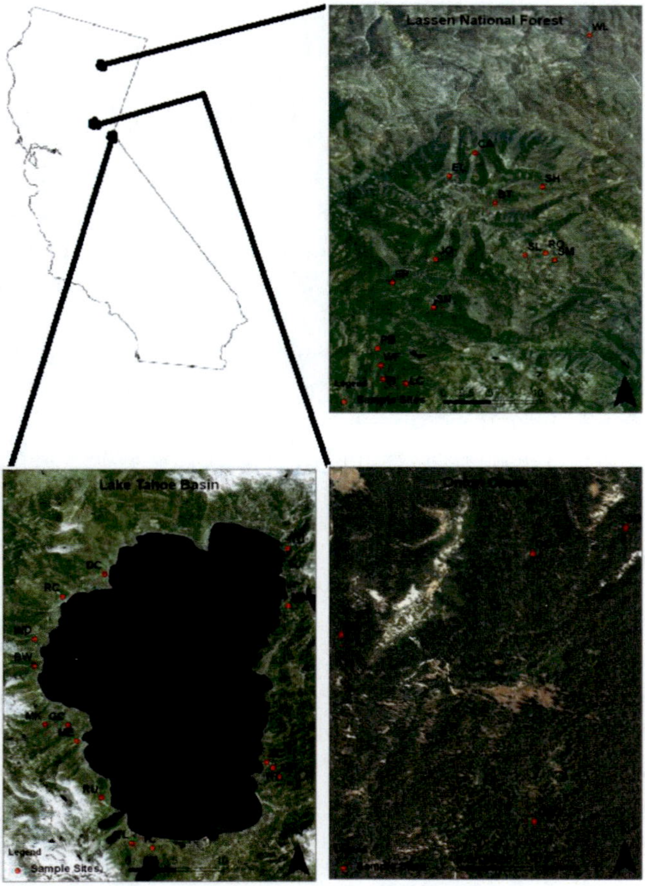

Figure 1. Location of the four sample areas in California (upper left), and of sample sites within each area.

Elevations ranged from 1519 m at Philbrook Creek on the Lassen National Forest to 2158 m at Tunnel Creek in the Lake Tahoe Basin. Longitudes ranged from 119° 55' W to 121° 30' W, and latitudes ranged from 38° 55' N to 40° 20' N. Most precipitation occurs during the winter as snow, and average annual totals (data from 1903 to 2009) varied from 460 mm on the east side of the Lake Tahoe Basin to 1340 mm on the Lassen National Forest (Beaty and Taylor, 2001; DRI, 2009). Forest composition varies widely with elevation, aspect and precipitation, and includes white fir (*Abies concolor*), red fir (*Abies magnifica*), Jeffrey pine (*Pinus jeffreyi*), ponderosa pine (*Pinus ponderosa*), sugar pine (*Pinus lambertiana*), lodgepole pine (*Pinus contorta ssp. murrayana*), western white pine (*Pinus monticola*), incense-cedar (*Calocedrus decurrens*), Douglas-fir (*Pseudotsuga menziesii*), black oak (*Quercus kelloggii*), quaking aspen (*Populus tremuloides*), black cottonwood (*Populus balsamifera ssp. trichocarpa*), mountain alder (*Alnus incana ssp. tenuifolia*), and willow (*Salix spp.*) in varying proportions. Jeffrey pine or ponderosa pine typically dominate on drier sites and south-facing slopes, while white fir or red fir typically dominate on wetter sites and north-facing slopes. Sampling was confined to Sierra Nevada forest types that were historically characterized by frequent (<30 year), low- to mixed-severity fire regimes.

Anthropogenic influence in all sampling areas has likely had a profound effect on stand structure and fuel loads. The Washoe Indians and their ancestors have inhabited the Lake Tahoe Basin for the last 8000 to 9000 years, and may have used fire to improve accessibility, wildlife habitat, hunting, and plant material quality. Major Euro-American settlement of the Tahoe Basin began when the first pack trail into the basin was completed in the 1850s. Logging began on the south shore of Lake Tahoe in 1859, and numerous settlements were established in the 1860s. Much of the Lake Tahoe Basin was heavily logged from the 1860s to 1890s to support the mining of Nevada's Comstock Lode. Accumulation of logging slash and introduction of new ignition sources such as sawmills, railroads, and logging equipment likely influenced fire frequency, residual stand structure, and fuel loads during this era (Lindstrom et al., 2000).

The Almanor Ranger District of the Lassen National Forest is located in Plumas County, which was also extensively logged beginning with the opening of the first sawmill in 1851 (Lawson and Elliot, 2008). The Onion Creek Experimental Forest was subject to considerably less human influence than the Tahoe Basin and Lassen areas, with only 20% of the area logged in the early 1900's (Berg, 1990). Because logging likely removed many of the larger trees with the longest tree ring records, sampling was concentrated on remnant late successional forest patches that would facilitate the best reconstruction of historic stand conditions.

Potential sites were identified by first consulting US Forest Service maps of late successional forest patches likely to contain a long fire record. Potential sites were then scouted to determine the prevalence of numerous fire-scarred trees, stumps, and logs. Sample sites were non-randomly chosen to provide a long record of fire history and to represent the variability of forest types and riparian area characteristics present in Sierra Nevada forests influenced by fire exclusion. Sites in some forest types and riparian zone width classes were greatly limited by fire scar availability, and sites with east side precipitation regimes were only available in the Lake Tahoe Basin sampling area (with the exception of Warner Creek in the Lassen National Forest sampling area). We found fifteen sites with Jeffrey pine forest type, seventeen sites with mixed-conifer forest type, and four sites with white fir forest type; twenty-nine sites with west side precipitation regimes, and seven sites with east side precipitation regimes.

Within these sample sites, plot locations were randomly selected for both the riparian and upland areas, with the stipulation that upland sites were located on the same side of the stream from which most historical fires likely approached, given local topography and regional wind patterns. This ensured that the effects of fire on stand structure and fuel loads measured in upland forests were not influenced by riparian microclimates. The riparian zone was determined by a combination of stream channel incision and understory plant community composition (i.e. riparian indicator species that were common throughout the study area, such as *Rubus parviflorus, Pteridium aquilinum, Alnus incana spp. tenuifolia, Salix spp.*). Riparian zone widths varied from 7 m on narrow ephemeral headwater streams to 420 m on wide alluvial flats of large perennial streams.

2.2. Plot-Level Data Collection

At each site, fire-scarred trees, stumps and logs were scouted and catalogued along both sides of an approximately 2 km length of stream with late-successional forest characteristics as identified on US Forest Service maps. During sampling, collection priority was given to specimens with a large number of fire events recorded (Swetnam and Baisan, 2003), large trees likely to have a long tree ring record, and dead material such as snags, stumps and logs with the least rotten wood. Full cross sections were obtained when possible from dead material, while partial cross sections were obtained from live trees in order to cause the least structural damage (Arno and Sneck, 1977). Eighteen to thirty-two specimens were collected at each site, taking roughly half from the upland area and half from the riparian area. For comparison, upland sample sites were selected ranging from 50 to 300 m from riparian sample sites, in the same forest community. Riparian and upland samples were collected over similar sized areas (within 20%). Forest, site and stream characteristics were recorded at each site, including forest type, percent species composition by basal area occupied by fire tolerant pine species (Jeffrey, ponderosa and sugar pine), precipitation regime (wet west side vs. dry east side), elevation, slope, aspect, stream order, riparian zone width, flow regime (perennial vs. intermittent), and channel bankfull width, depth, width to depth ratio, gradient, and approximate shape. We defined first order streams as the smallest streams visible on standard USGS 7.5 minute topographic maps, second order streams as those formed by the junction of two first order streams, etc. (Strahler, 1952).

Forest structure, species composition, and fuel loads were measured in each of the 36 adjacent riparian and upland sites (72 plots total). Paired riparian and upland plots were each randomly placed in the general locations where fire scar samples had been collected for analysis of fire history. Riparian plots were placed parallel to the channel with the streamside edge located at the bankfull stage, defined as the highest position on the stream bank reached by flows of a 1.5 year recurrence interval, and often identifiable by a change in bank steepness and vegetation composition (Dunne and Leopold 1978). Upland plots were generally located 50 to 300 m away in the same watershed and forest type to ensure adequate pairing with corresponding riparian plots. Plots were rectangular, with a 0.05 ha plot (25 m long by 20 m wide) nested inside a 0.1 ha plot (50 m long by 20 m wide). Where riparian zones were less than 20 m wide on each side of the channel, riparian plot dimensions were adjusted accordingly to ensure accommodation of the entire plot area within the riparian zone.

Species, structure type (live, snag, log, stump), diameter at breast height (DBH), total height, and height to the first live branch were measured for all structures larger than 5 cm DBH within a 0.05 ha rectangular plot, and for all trees larger than 50 cm DBH within a 0.1 ha rectangular plot. Snags were defined as any standing dead trees greater than 1.3 m in height that were not in full contact with the ground, logs were defined as any fallen trees in full contact with the ground and attached to a root wad originating within the plot, and stumps were defined to be less than 1.3 m in height. Additionally, between 7 and 22 trees of representative diameter classes and species were cored to the pith at soil height within each plot to aid in stand reconstruction. Representative diameter classes varied with the range of diameters present at each site, but generally consisted of trees 20-50 cm, 50-80 cm, and >80 cm DBH. Surface fuel loads (dead and down woody material, litter and duff) were measured along 3 transects at each plot using Brown's planar intercept method (Brown, 1974). Shrub cover by species and canopy cover were measured along two 50 m transects along the sides of each plot, and seedling species and height were recorded in ten 1 m^2 plots located at 10 m intervals along the same transects.

2.3. Reconstruction Methods

The reconstruction period for each plot was set at the year of the last fire at that site, which ranged from 1848 to 1990 as determined from site-specific fire scar records, with 64% of the periods before 1940. Mean (standard deviation) reconstruction periods for the Lassen, Onion Creek, east and west Tahoe sampling areas were 80 (41), 116 (13), 98 (39), and 78 (42) years before 2009 for riparian plots, and 72 (35), 104 (39), 92 (32), and 91 (40) years before 2009 for upland plots. Historical stand structure at the time of the last fire was reconstructed for each riparian and upland plot following methods commonly used in southwestern ponderosa pine forests, in which the DBH, total height, and height to the first live branch of live trees during the reconstruction period are estimated from measurements of current live trees, snags, logs, and stumps (Fulé et al., 1997; Mast et al., 1999; Moore et al., 2004).

These reconstruction methods were adapted to accommodate the presence of both shade-intolerant and shade-tolerant species in Sierra Nevada mixed-conifer forest types. Because the cores taken from each plot (761 total) provided a sample proportional to the current species composition, the most frequent shade-tolerant species such as white fir, incense-cedar, and red fir comprised the majority of the sample. This favored representation of the greater variability in annual radial increment commonly exhibited in shade-tolerant species (Oliver and Larson, 1996). The 761 cores were sanded with 400 grit sandpaper to allow accurate visual identification of tree ring boundaries, and were manually cross-dated using standard procedures (Stokes and Smiley, 1968).

For trees with stem rot or large diameter that could not be cored to the pith (26% of all cores taken), methods following Scholl and Taylor (2010) were used to determine the ages of trees with incomplete cores. Regressions between DBH and core length inside bark were developed for each species from complete cores. All regressions were significant (p values ranging from <0.001 to 0.016) with r^2 values ranging from 0.477 to 0.876. For incomplete cores, actual core length was subtracted from predicted core length to determine the missing length. For cores in which the predicted length was less than the actual length, the actual length was used as the closest approximation for the total core length (Scholl and Taylor, 2010). From the complete cores, the average number of rings per centimeter for each species was determined from the width of the first five years' growth. The number of rings per centimeter was multiplied by the missing length of incomplete cores to determine the number of missing years from the end of the core to the pith. Tree age was then estimated by adding the missing years to the incomplete core age.

2.3.1. Live Tree Reconstruction

To reconstruct the historic size of current live trees measured in each plot, species-specific equations were used to predict current DBH inside bark from current DBH outside bark (Dolph, 1981). Reconstructed DBH inside bark (at the time of the last fire in each site) was estimated using mean annual radial increment calculations for each species developed from regression equations predicting DBH inside bark from tree age using data from the cores. Regressions equations for each species were applied to trees of that species in all four sampling areas. All regressions were significant ($p<0.001$), with r^2 values ranging from 0.872 to 0.984. Reconstructed DBH outside bark was then estimated using the equations in Dolph (1981).

Reconstructed total tree height and height to the first live branch were estimated using regression equations (developed from current tree data) for each species predicting those variables from DBH. Although trees under historical conditions of low stand density likely had crown structure different from that observed under current conditions, historical crown structure data sufficient for developing regression equations was unavailable. All regressions were significant (p values <0.001 for height equations, and ranging from <0.001 to 0.050 for height to first live branch equations), with r^2 values ranging from 0.870 to 0.991 for height, and 0.501 to 0.861 for height to first live branch. Reconstructed stand-level characteristics were then calculated from the reconstructed DBH, total height, and height to the first live branch of live trees (Table 2).

2.3.2. Snag and Stump Reconstruction

Reconstructing the status and size of snags, logs, and stumps at the time of the last fire from increment cores is often not feasible because the extensive rot present in many structures makes determining the year of death impossible. Year of tree death for snags in each plot was estimated from a field rating of decay class and species-specific equations predicting decay class transition times (Morrison and Raphael, 1993; Raphael and Morrison, 1987). The live tree equations were then applied to reconstruct tree size at the time of the last fire for snags that died after the reconstruction year. Reconstructed snag volume was calculated from snag DBH using published species-specific volume equations (Table 2) (Wensel and Olson 1995). The year of stump death was determined from field observations of stump characteristics and known logging periods (Lawson and Elliot, 2008; Lindstrom et al., 2000), and the live-tree equations were applied to estimate tree size at the time of the last fire for trees that were cut after the reconstruction year.

Table 2. Reconstruction approaches for stand-level characteristics.

Stand Metric	Reconstruction Approach
BA (m^2/ha)	Individual live tree BA calculated from DBH, expanded to hectare scale and summed
Stand Dens (stems/ha)	Individual live tree counts expanded to hectare scale and summed
QMD (cm)	Square root taken of the sum of live tree diameters squared divided by the number of live trees in the plot
Avg CBH (m)	Height to first live branch calculated for individual trees using species-specific regression equations, then averaged
% Comp (by BA)	BA summed for fire tolerant and intolerant species groups (Table 3), then divided by total BA
Snag Vol (snags/ha)	Individual snag volumes calculated from DBH using published equations, expanded to hectare scale and summed
Fuel Loads (Mg/ha)	Published fuel deposition rates applied to # yrs elapsed between last and second-to-last fires at each site

2.3.3. Log Reconstruction

The year of transition from snag to log for logs in each plot was estimated using species-specific equations predicting log age from field rating of decay class (Kimmey, 1955; Harmon et al., 1987). Year of tree death for snags that subsequently transitioned to logs was estimated as described above using snag decay class transition time equations. Tree size at the time of the last fire was then estimated for logs that originated from snags that died after the reconstruction year, using the live tree equations as described earlier. Tip-up mounds and consistent log orientation with prevailing winds were rare, indicating that most down trees were snags prior to becoming logs, a trend noted elsewhere in Sierran mixed-conifer forests (Innes et al., 2006; North et al., 2007). However, this method for reconstructing live-tree DBH from log diameter and decay class should be considered a conservative estimate for any trees that were blown down by wind events without first dying and becoming snags.

2.3.4. Forest Structure Reconstruction Limitations

Forest reconstruction is limited by the material currently existing on site, which may influence estimation of historical stand conditions. For example, the east side of the Tahoe Basin receives the least precipitation of the sampling areas and thus has numerous well-preserved stumps from 19th century logging which facilitate highly accurate forest reconstructions (i.e. Taylor, 2004). However, the sites in Lassen, Onion Creek and the west side of the Tahoe Basin receive higher precipitation and have faster decay rates, resulting in fewer snags, stumps and logs with intact tree ring records, which likely reduces reconstruction accuracy. These reconstruction methods are also extremely limited in their ability to estimate historic density of small trees (North et al., 2007). Current snags, logs, and stumps measured in each plot would under-represent small trees that died after the reconstruction period and fully decayed before 2009, especially those species with rapid decay rates such as white fir (Harmon et al., 1987; Morrison and Raphael, 1993). Reconstruction estimates of small tree density are likely to be conservative, thus affecting estimated stand density, basal area (BA), quadratic mean diameter (QMD), canopy base height (CBH), and species composition at the time of the last fire.

2.3.5. Fuel Load Reconstruction

Fuel loads for reconstructed stands were estimated using published species- and size-specific equations for deposition rates of different fuel classes (van Wagtendonk and Moore, 2010). These rates were applied to site-specific fuel accumulation times for each site, defined as the number of years between the last fire and the second-to-last fire, to obtain reconstructed fuel loads by fuel size class (Table 2). The limitations of reconstructing fuel loads vary by size class, with larger size classes (1000 hr fuels) being more prone to error than smaller size classes (duff, litter, 1-100 hr fuels). This can be attributed to the greater difficulty in sampling the more variable deposition rates of coarse woody debris produced largely by episodic tree mortality, relative to the more regular input of fine woody and herbaceous fuels from litterfall (Keane, 2008).

2.3.6. Fire Behavior and Effects Modeling

The current and reconstructed fuel loads, stand structure, and species composition were entered into the Forest Vegetation Simulator (FVS) to produce stand visualizations, and run through the Fire and Fuels Extension (FFE) to model potential fire behavior, effects, and canopy bulk density (Dixon, 2002; Reinhardt and Crookston, 2003; Wykoff et al., 1982). FVS is a regionally calibrated growth and yield model that can, among numerous other applications, produce accurate visualizations of stand structure and species composition from plot-level data inputs such as status (live or dead), species, DBH, height, and crown ratio. The Western Sierra Nevada variant was used for Tahoe and Onion Creek sites, while the Inland California and Southern Cascades Variant was used for the Lassen sites (Dixon, 2002). The FFE links the stand structure and species composition data to plot-level fuel load data (duff, litter, 1-1000 hr), local fuel moisture data (1-1000 hr dead woody, duff, live woody and herbaceous), and weather data (temperature, relative humidity, wind) to produce estimates of potential fire behavior and effects (Reinhardt and Crookston, 2003).

The fuel moisture and weather inputs used to model potential fire behavior in FVS-FFE were 97th percentile conditions (Table 3) from historical data gathered by representative remote automated weather stations at Chester, CA (for Lassen sites) and Meyers, CA (for Tahoe and Onion Creek sites). Data was obtained from the National Interagency Fire Management Integrated Database via the Kansas City Fire Access Software, and 97th percentile conditions were calculated using FireFamily Plus software (Bradshaw and McCormick, 2000; U.S. Forest Service, 1993; U.S. Forest Service, 1996).

Table 3. 97th percentile fuel moisture and weather conditions used to model potential fire behavior.

	Fuel Moisture (%)						Weather Conditions		
	Dead Woody				Live		6.1 m Wind	RH	Temp
	1 hr	10 hr	100 hr	1000 hr	Wood	Herb	(km/h)	(%)	(°C)
Tahoe	2	3	6	7	59		10	9	31
Lassen	2	3	6	7	69		19	7	34

2.4. Statistical Analyses

2.4.1. Fire History

Specimens were sanded using progressively finer sand paper, and fire scars were dated using standard crossdating procedures (Stokes and Smiley, 1977). The season of each fire event was determined from intra-annual ring position, and classified as occurring in earlywood (early, middle or late), latewood, or the dormant season after cessation of tree growth (Dieterich and Swetnam, 1984). Analysis of fire scar data was performed using FHX2 software (Grissino-Mayer, 2001). Fire return intervals were analyzed using two composite filters: a broad filter (C1) including every fire event recorded on every specimen to provide a liberal estimate of fire return interval, and a narrow filter (C10) including only fire events recorded on two or more specimens at a given site (about 10% of the specimens) to provide a more conservative estimate of fire return interval based on larger fire events (Swetnam and Baisan, 1996). The time period of fire events recorded in the specimens varied by site, and fire return interval analysis was restricted to time periods beginning when a fire event was recorded by two or more specimens at a given site. Percent dormant season fire scars, and mean C1 and C10 fire return intervals of adjacent riparian and upland sites were compared using a paired t test. Riparian and upland C1 fire return intervals before and after 1850, including the last incomplete interval in sites where fire was absent after 1850, were compared using a paired t test. An alpha value of 0.1 was used to determine statistical significance, due to the high level of spatial and temporal variability in fire return intervals.

Nonmetric multidimensional scaling (NMS), performed in PCORD (MjM Software Design; McCune and Grace, 2002), was used to assess forest, site and stream characteristics influencing riparian and upland fire return intervals under the C1 and C10 filters. Variables with highly skewed distributions were log-transformed accounting for the lowest non-zero value (McCune and Grace, 2002). Because different measurements are subject to different scales, all variables were relativized by adjusting values to the standard deviation of each variable's mean value. Outlier sites were assessed for their potential influence on the ordination, but none were removed from the analysis because none were >4 standard deviations from the mean, and all were considered to have important fire regime information. Because not all sites had a fire scar record sufficient to calculate all C1 and C10 fire return interval metrics used in the analysis (mean, median, Weibull modal, Weibull median, minimum, maximum), the ordination matrices were reduced to 34 sites for riparian C1 FRI, 20 sites for riparian C10 FRI, 35 sites for upland C1 FRI, and 25 sites for upland C10 FRI. NMS was run using the Sorenson distance measure, 4 starting axes, 15 iterations, and an instability criterion of 0.0001. A joint plot of significant (r^2>0.1) forest, site and stream variables was overlaid on the ordination of sites and fire return intervals.

Regression trees, a component of classification and regression tree analysis (CART) in S-Plus (Breiman et al., 1984; Moore et al., 1991), were used to further investigate forest, site and stream characteristics associated with the variability in C1 and C10 mean fire return intervals among riparian and upland sites. Regression tree analysis is a nonparametric, recursive model well suited to exploring ecological relationships that are difficult to detect using other multivariate analyses (De'ath, 2002; Vayssieres et al., 2000). Each regression tree was pruned to a minimum of 10 observations before a split, a minimum node size of 5 and a minimum node deviance of 0.5. Because not all sites had a fire scar record sufficient to calculate C10 FRI, the riparian and upland C10 FRI regression trees were restricted to 28 and 31 sites, respectively.

For investigation of fire-climate synchrony, we restricted our analysis to years in which two or more specimens were scarred at a site, and two or more sites recorded fire scars in the same year (Dieterich, 1980). Because most of our sites were widely separated, scars at different sites were not assumed to be produced by the same fire event, but separate fire events favored by climate conditions in the same year. Superposed epoch analysis (SEA) was used to compare climate six years before, the year of, and four years after each year in which fire occurred at two or more sites (Swetnam, 1993). We used data from the reconstructed Palmer Drought Severity Index (PDSI) gridpoint 46 (closest to our sample locations) and the NINO3 index, which represents the mid-tropical Pacific sea surface temperature fluctuations associated with the El Niño/Southern Oscillation (ENSO) (Cook, 2000; Cook et al., 1999). PDSI has been correlated with fire events on the west side of the Sierra Nevada (Norman and Taylor, 2003; Stephens and Collins, 2004; Swetnam and Baisan, 2003; Taylor and Beaty, 2005), while ENSO has been correlated with fire events in the Southwest (Grissino-Mayer and Swetnam, 2000; Skinner et al., 2008), the Sierra Nevada (Beaty and Taylor, 2008; Norman and Taylor, 2003), and the Pacific Northwest (Heyerdahl et al., 2008; Kitzberger et al., 2007). Values for both indices were standardized around a mean of zero, and 95% confidence intervals were calculated using Monte Carlo simulations with 1000 iterations.

2.4.2. Stand Structure, Fuel Loads and Fire Behavior

Measures of stand structure (BA, stand density, snag volume, QMD, average CBH), species composition (by percent of total BA, categorized fire-tolerant and -intolerant functional groups), fuel load (by duff, litter, 1 hr, 10 hr, 100 hr, 1000 hr classes), potential fire behavior (surface and crown fire flame lengths, probability of torching, torching index, crowning index), canopy bulk density (CBD), and mortality (by percent of total BA) were compared between current and reconstructed riparian and upland forests, as well as between sampling areas, using ANOVA. Variables were checked for normality of residuals using normal probability plots and the Shapiro-Wilk test (Shapiro and Wilk, 1965), and all variables except BA, QMD, CBH, and probability of torching were determined to have residual distributions with significant departures from normality. Homogeneity of variances was assessed using plots of residual vs. predicted values, and residual values vs. fixed factors, and the variances of all variables were determined to be heterogeneous. Various logarithmic, power, and arcsine transformations were applied to improve normality of residuals and heterogeneity of variances. Normality of residuals was achieved for all variables except species composition, and homogeneity of variances was achieved for all variables except snag volume, surface and crown fire flame length, probability of torching, and mortality. Results involving these variables should be treated with caution.

The experiment set up as a split plot repeated measures design, with site as the main plot, riparian vs. upland as the split plot, and current vs. reconstructed as a repeated measure. A linear mixed effects model was used to analyze the data, which included sampling area, riparian vs. upland, and current vs. reconstructed as fixed effects, and site and plot as random effects. All possible interactions were included in the model, and differences between the least squares means of current riparian, reconstructed riparian, current upland, and reconstructed upland variables, as well as differences among sampling areas, were compared using a Tukey's post-hoc test of the riparian vs. upland by current vs. reconstructed interaction, and of the sampling area factor.

In this study we were also interested in how fire history might affect reconstructed stand and fuel conditions, and whether riparian and upland conditions were correlated. To explore the relationships between site-specific broad filter fire return intervals (C1 FRI, derived from all fire events scarring one or more trees at a given site) and reconstructed stand structure, fuel loads, and potential fire behavior, we checked variables for normality and then used a Pearson's correlation coefficient analysis. Separate correlation matrices were set up for riparian and upland forests, with reconstructed stand structure, fuel loading and fire behavior variables correlated with FRI. A Pearson's correlation coefficient analysis was also used to examine the relationships between riparian and upland forests for both current and reconstructed conditions. Separate correlation matrices were set up for current and reconstructed conditions, with riparian stand structure, fuel loading and fire behavior variables correlated with the corresponding upland variables. ANOVAs were performed using SAS software, Version 9.1.3 of the SAS System for Windows, Copyright © 1998 SAS Institute Inc., while correlation analyses were conducted using Minitab Version 16 (McKenzie and Goldman, 1999).

One reconstructed riparian plot (Taylor Creek on the west side of the Tahoe Basin) had so few trees that torching index, crowning index, and canopy bulk density could not be calculated in FVS-FFE, resulting in a sample size of 35 for those variables in the reconstructed riparian plot category. One reconstructed upland plot (Burke Creek on the west side of Tahoe Basin) had no trees detectable by the reconstruction methods used, resulting in a sample size of 35 for all variables in the reconstructed upland plot category.

3. RESULTS

3.1 Riparian vs. Upland FRI and Seasonality

In total, 907 fire scar specimens were collected from 36 sites. The analysis included 849 specimens (58 could not be crossdated) with 1631 fire scars recording 760 independent fire events.

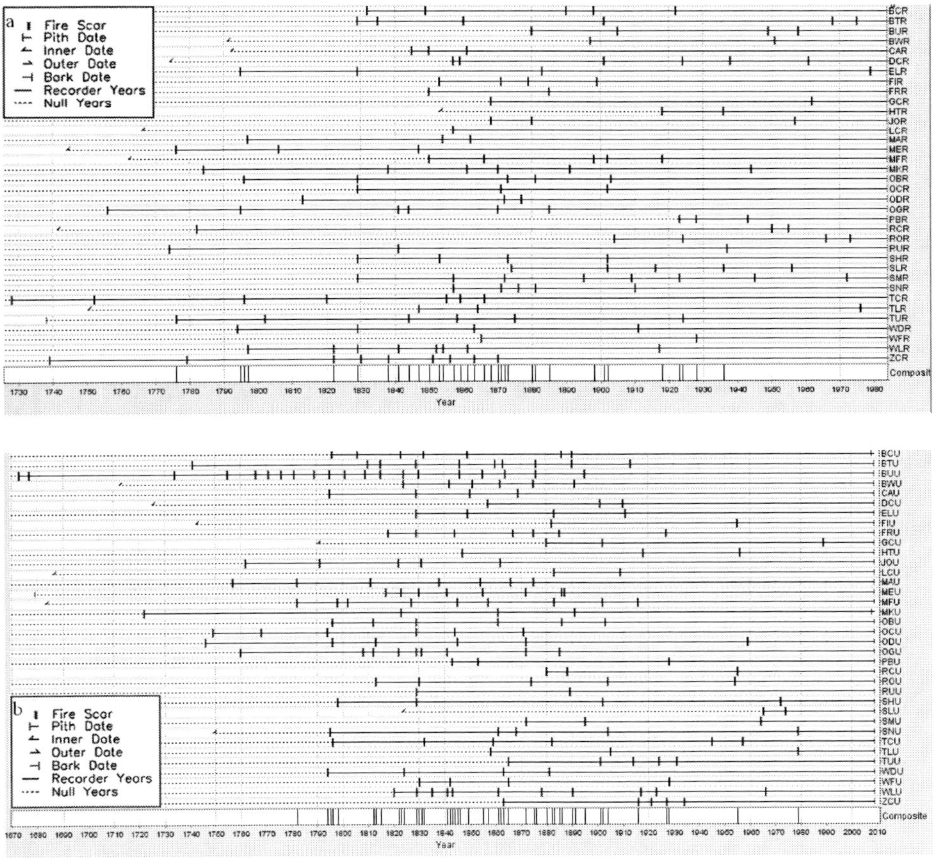

Figure 2. Composite fire occurrences for 36 sample sites in a) riparian and b) adjacent upland forests. Horizontal lines are the length of record at each sample site, with vertical ticks indicating years when two or more trees were scarred at a site. The composite record at the bottom indicates when two or more sites were scarred in the same year.

Riparian fire scar samples, which were predominantly true fir and incense-cedar, exhibited more complacent ring series than upland samples, and were thus moderately more difficult to cross-date. This was likely due to a combination of higher soil moisture and a higher proportion of true firs and incense-cedars in riparian areas. The period of record ranged from 1387, the earliest ring on a Jeffrey pine stump from Burke Creek in the Lake Tahoe Basin, to 2009, the year sampling took place. The earliest fire event recorded was 1526 on a Jeffrey pine snag from Taylor Creek in the Lake Tahoe Basin, and the latest fire event recorded was 2005 on a live lodgepole pine from Tallac Creek in the Lake Tahoe Basin.

Riparian mean C1 FRI ranged from 8.4 years in Taylor Creek to 42.3 years in Blackwood Creek, both Jeffrey pine sites in the Tahoe sampling area. Upland mean C1 FRI varied from 6.1 years in Burke Creek, a Jeffrey pine site in the Tahoe sampling area, to 58.0 years in Shanghai Creek, a mixed-conifer site in the Lassen sampling area (Table 1). Across all sites, the average C1 FRI in the riparian and upland areas was 16.6 and 16.9 years, respectively. Riparian mean C10 FRI varied from 10.0 years in Philbrook Creek, a mixed-conifer site in the Lassen sampling area, to 86.5 years in the Red Cedar Creek, a mixed-conifer site in the Tahoe sampling area (Figure 2a). Upland mean C10 FRI varied from 10.0 years in Meeks Creek, a mixed-conifer site in the Tahoe sampling area, to 56.3 years in McKinney Creek, a Jeffrey pine site in the Tahoe sampling area (Figure 2b). Across all sites, the average C10 FRI in the riparian and upland areas was 30.0 and 27.8 years, respectively. The C10 FRI could not be calculated for 11 sites due to insufficient fire scar records.

Table 1. Forest, site and stream characteristics; forest type abbreviations are Jeffrey pine (PIJE), white fir (ABCO) and mixed-conifer (MC); C1 FRI (all fire events at a site after two trees are scarred), C10 FRI (fire events scarring two or more trees at a site) for each riparian and upland site, and p-values of paired t tests (* and ** indicate significance at α=0.1 and 0.05, respectively)

Area	Site	Forest Type	Precip Regime	Elev (m)	Stream Order	Riparian Width (m)	Stream Width (m)	Stream Depth (m)	Channel Grad (°)	Area Sampled (ha) Riparian	Area Sampled (ha) Upland	C1 Mean FRI Rip	C1 Mean FRI Up	C1 Mean FRI p Diff	C10 Mean FRI Rip	C10 Mean FRI Up	C10 Mean FRI p Diff
Lassen National Forest	Butt (BT)	PIJE	W	1700	1	8	4.0	0.6	4	2	4	19.2	8.8	**0.0081	24.3	19.1	0.5806
	Carter (CA)	MC	W	1767	2	20	1.0	1.0	1	1	1	9.5	13.8	*0.0759	NA	24.7	NA
	Elam (EL)	MC	W	1746	2	21	6.0	2.0	7	1	1	21.4	16.5	0.7982	61.3	27.3	*0.0976
	Fish (FI)	MC	W	1548	2	70	10.2	1.2	3	1	2	12.5	12.3	0.6738	15.3	28.0	0.8949
	W Feather (FR)	MC	W	1536	3	35	25.0	5.0	5	1	2	14.5	13.9	0.8145	NA	18.2	NA
	Jones (JO)	PIJE	W	1623	2	12	12.0	0.8	2	1	2	17.7	13.0	0.1254	44.5	12.2	0.4908
	Last Chance (LC)	PIJE	W	1648	2	37	11.0	2.7	4	1	1	16.0	12.0	0.2319	NA	NA	NA
	Philbrook (PB)	MC	W	1531	3	35	35.0	10.0	8	1	1	13.0	42.5	0.2139	10.0	42.5	0.4255
	Rock (RO)	ABCO	W	1846	2	20	15.0	2.0	3	1	2	21.0	14.8	*0.076	23.0	35.3	NA
	Shanghai (SH)	MC	W	1694	2	16	10.0	1.1	6	3	5	15.7	58.0	**0.0105	24.3	52.0	0.1512
	Slate (SL)	ABCO	W	1900	1	20	20.0	1.5	2	3	3	20.5	11.8	**0.0499	20.5	NA	NA
	Sawmill Tom (SM)	ABCO	W	1754	2	15	12.0	2.0	6	3	3	11.9	13.5	0.5841	17.9	46.0	*0.0616
	Snag (SN)	ABCO	W	1724	2	85	13.0	1.1	2	2	1	15.9	23.0	0.5228	17.7	46.0	0.1935
	W Fish (WF)	MC	W	1557	2	20	15.0	2.0	4	2	1	25.7	31.8	0.6952	NA	32.7	NA
	Warner (WL)	PIJE	E	1543	4	25	9.5	2.0	1	7	7	8.5	10.8	0.3927	17.1	14.6	0.9216
Onion Creek	Onion B (OB)	MC	W	1954	1	11	9.0	1.7	7	1	2	22.2	15.8	0.3292	35.7	21.4	**0.0332
	Onion C (OC)	MC	W	1907	1	35	9.7	1.4	5	4	7	15.2	13.4	0.6314	36.5	20.3	*0.059
	Onion D (OD)	MC	W	1997	1	40	24.0	2.0	4	7	8	20.3	24.8	0.3551	32.0	42.6	0.6064
	Onion G (OG)	MC	W	1908	1	19	4.7	1.5	9	2	3	12.0	7.4	*0.0815	25.8	15.6	0.3051

Table 1. (Continued)

	Bunker (BC)	MC	W	1972	1	45	0.8	0.2	8	1	1	20.2	13.8	0.6825	22.5	23.5	0.9159
	Burke (BU)	PIJE	E	2053	1	7	7.6	0.3	3	2	2	14.3	6.1	**0.0002	26.0	11.1	**0.0382
	Blackwood (BW)	PIJE	W	1868	2	75	14.0	1.4	2	15	12	42.3	19.4	*0.069	NA	13.4	NA
	Dollar (DC)	PIJE	W	1948	1	35	1.7	0.9	3	2	2	14.4	13.7	0.9708	20.8	26.5	0.6233
	General (GC)	MC	W	1951	3	163	14.5	1.5	1	3	5	16.0	17.4	0.7484	56.0	54.6	NA
	Horse Trail (HT)	PIJE	E	1952	1	37	1.8	1.3	1	1	1	9.0	10.7	0.8693	NA	NA	NA
	Marlette (MA)	PIJE	E	1941	2	23	3.0	0.9	2	2	2	22.1	16.9	0.3055	32.5	21.3	0.6818
Lake Tahoe Basin	Meeks (ME)	MC	W	1902	2	420	6.0	1.0	1	2	2	15.6	11.7	0.3835	35.5	10.0	*0.0519
	McFaul (MF)	PIJE	E	2042	2	75	1.4	0.6	1	3	2	10.7	10.9	0.9594	17.0	16.8	0.8423
	McKinney (MK)	PIJE	W	1989	1	30	12.0	2.0	3	3	4	10.0	11.4	0.3127	32.0	56.3	0.2774
	Red Cedar (RC)	MC	W	1948	1	40	1.7	0.3	2	2	1	19.6	17.6	0.8351	86.5	37.5	0.5851
	Rubicon (RU)	MC	W	2065	1	26	1.0	0.2	5	2	2	14.8	29.3	0.1208	NA	NA	NA
	Taylor (TC)	PIJE	W	1914	3	86	23.0	4.0	1	10	10	8.4	15.1	0.1046	23.0	32.2	0.3540
	Tallac (TL)	MC	W	1901	2	355	5.0	0.6	1	12	13	16.1	11.6	0.4498	NA	NA	NA
	Tunnel (TU)	PIJE	E	1976	1	10	4.0	1.6	4	14	12	16.8	11.6	*0.0895	29.6	16.5	0.1792
	Ward (WD)	PIJE	W	1944	2	34	18.0	1.5	3	1	1	22.8	18.7	0.6126	39.0	29.0	0.2555
	Zephyr (ZC)	PIJE	E	1831	3	26	23.6	0.6	3	1	1	12.0	13.4	0.6504	14.6	14.2	0.8406

C1 FRI was significantly different between riparian and upland areas in only 9 out of 36 sites, with riparian FRI being shorter than upland FRI in two of these sites (Carter and Shanghai Creeks, both mixed-conifer sites in the Lassen sampling area) (Table 1). C10 FRI was significantly different between riparian and upland areas in only 6 out of 25 sites, with riparian FRI being shorter than upland FRI in one of these sites (Sawmill Tom Creek, a white fir site in the Lassen sampling area).

Fire seasonality varied by site but with the exception of six sites (Burke, Dollar, and Horse Trail Creeks, the riparian area of Bunker Creek, and the upland areas of Meeks and Zephyr Creeks), >50% of the scars were in the dormant season. Averaged across all sites, 88% and 79% of scars in the riparian and upland areas, respectively, were in the dormant season (Figure 3). There was no significant difference in the percent dormant season scars between the riparian and upland areas (p=0.102).

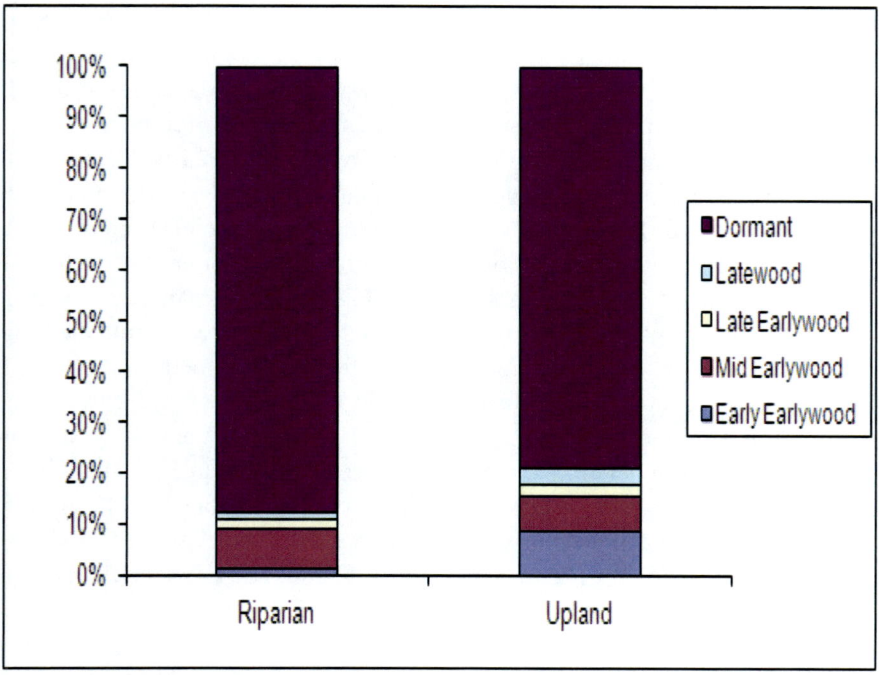

Figure 3. Cumulative proportion of intra-ring fire scar positions for all trees across all sites. Dormant typically represents fall and late summer fires, latewood represents mid-summer fires, earlywood represents spring and early summer fires.

3.2. Site Characteristics Associated with Riparian and Upland FRI

For the NMS analyses, the greatest reduction in stress was achieved with two axes. In the riparian C1 FRI analysis, the proportion of variance (the fit between distance in the ordination space and the original space) represented by the first and second axes was 0.572 and 0.385, respectively (cumulative 0.957). The joint plot of riparian C1 FRI (Figure 4a) shows a trend of decreasing FRI with decreasing channel width to depth ratio and increasing upland percent species composition occupied by pine. In the riparian C10 FRI analysis, the proportion of variance represented by the first and second axes was 0.133 and 0.827, respectively (cumulative 0.960). The joint plot of riparian C10 FRI (Figure 4b) reveals a trend of decreasing FRI with decreasing riparian zone width, decreasing channel gradient, and increasing upland and riparian percent species composition occupied by pine. In the upland C1 FRI analysis, the proportion of variance represented by the first and second axes was 0.359 and 0.629, respectively (cumulative 0.988). The joint plot of upland C1 FRI (Figure 4c) shows a trend of decreasing FRI with increasing upland percent species composition occupied by pine. In the upland C10 FRI analysis, the proportion of variance represented by the first and second axes was 0.865 and 0.117, respectively (cumulative 0.982). The joint plot of upland C10 FRI (Figure 4d) reveals a trend of decreasing FRI with increasing elevation and increasing upland percent species composition occupied by pine. All four NMS ordinations show a general clustering of east side sites in the quadrant of the ordination space where FRI is shorter.

In the riparian C1 FRI regression tree (Figure 5a), the shortest mean FRI (13.9 years) is associated with channel width to depth ratio <6.2, while the longest mean FRI (22.9 years) is associated with bankfull width to depth ratio >6.2 and channel bankfull depth >1.3 m. In the riparian C10 FRI regression tree (Figure 5b), the shortest mean FRI (23.9 years) is associated with mixed-conifer forest type and incised channel shape, closely followed by a mean FRI of 24.9 years associated with Jeffrey pine and white fir forest types. The longest C10 mean FRI (45.9 years) is associated with mixed-conifer forest type, and broad and v-shaped channels. In the upland C1 FRI regression tree (Figure 5c), the shortest mean FRI (13.1 years) is associated with >22.7% upland species composition occupied by pine, while the longest mean FRI (31.5) is associated with <22.7% upland percent species composition occupied by pine and elevation <1709 m. In the upland C10 FRI regression tree (Figure 5d), the shortest mean FRI (19.5 years) is associated with >37.6% upland species composition occupied by pine, while the longest mean FRI (42.9 years) is associated with <37.6% upland species composition occupied by pine and elevation >1944 m.

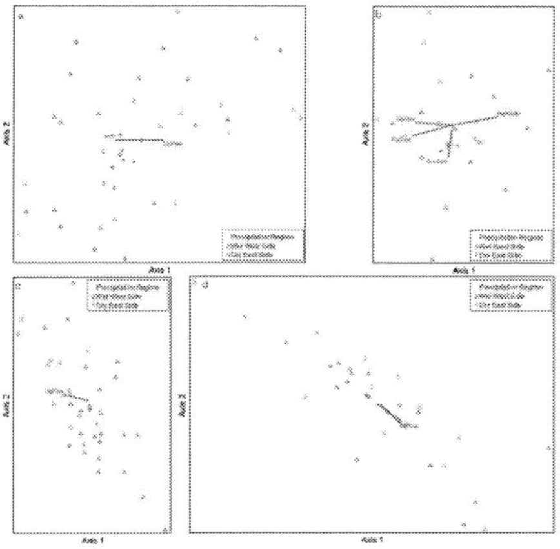

Figure 4. Nonmetric multidimensional scaling (NMS) ordination of a) riparian C1 FRI metrics at 34 sites (two sites not included due to insufficient fire scar record, i.e. site did not record enough fire events to allow calculation of all fire return interval metrics used in analysis); b) riparian C10 FRI metrics at 20 sites (16 sites not included due to insufficient fire scar record); c) upland C1 FRI at 35 sites (one site not included due to insufficient fire scar record); and d) upland C10 FRI at 25 sites (11 sites not included due to insufficient fire scar record). Triangle color represents precipitation regime of each site. The joint plots show the significant forest, site and stream characteristics associated with each fire occurrence record. Abbreviations: W/D is channel width/depth ratio, UpPine and RipPine are the upland and riparian percent species composition occupied by fire-tolerant pine, respectively, RipWidth is riparian zone width, and Gradient is channel gradient.

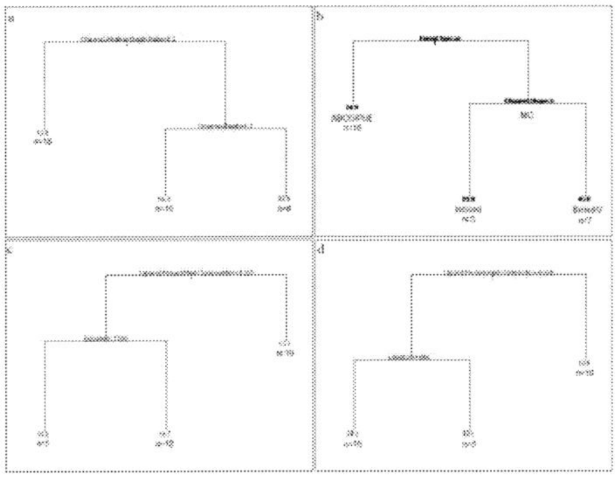

Figure 5. Regression tree of forest, site and stream characteristics associated with a) riparian C1 FRI, b) riparian C10 FRI, c) upland C1 FRI, and d) upland C10 FRI. The grouping of values in each split is indicated by the direction of the < symbol (i.e. in 5a, sites with channel depth <1.3 m are split off to the left of the dendrogram in the second split). The length of each branch is proportional to the amount of data variability explained by each split. Terminal values are the average FRI for all the sites classified in that node.

3.3. Climate Comparisons

SEA of years in which fires scarring two or more trees at a site occurred at two or more sites (the composite in Figures 2a and 2b) revealed a common pattern of fire-climate synchrony between the riparian and upland areas. SEA with PDSI identified a significant association between fire events and drought in the same year, but not in pre- or post-event years, in both riparian and upland areas (Figures 6a and 6b). There was a greater departure from mean PDSI in the upland areas (beyond the 99% confidence interval) than in the riparian areas (beyond the 95% confidence interval). SEA with the NINO3 index showed no significant associations between fire events and climate (not shown).

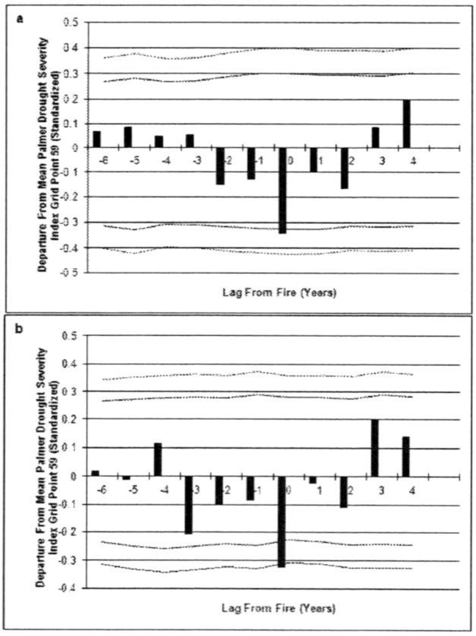

Figure 6. Superposed epoch analysis (SEA) for a) riparian and b) upland samples. Graphs show departure from the mean Palmer Drought Severity Index (PDSI) values with years when fires scarred two or more trees per site at the riparian areas of two or more sites. Horizontal lines are 95% and 99% confidence intervals.

3.4. Temporal Analysis

Temporal variation in FRI could not be analyzed for 10 riparian sites and four upland sites due to a lack of recorded fire occurrence before 1850 in those sites. C1 FRI was significantly different in two out of 26 riparian sites (Rubicon Creek, a mixed conifer site, and Taylor Creek, a Jeffrey pine site, both in the Tahoe sampling area), and two out of 32 upland sites (Butt Creek, a Jeffrey pine site, and Fish Creek, a mixed conifer site, both in the Lassen sampling area). In all four sites, C1 FRI was significantly shorter after 1850.

3.5. Comparison between Riparian/Upland Current/Reconstructed Stand Conditions

While there was a great deal of variability in riparian and upland forests under current and reconstructed conditions (Figure 7), the analysis revealed some striking differences. Current riparian forest conditions significantly differed from reconstructed riparian conditions in BA, stand density, snag volume, duff, 1 hr, 10 hr, 100 hr, and total fuel loads, surface and crown fire flame length, probability of torching, torching index, crowning index, CBD, and mortality (Table 4).

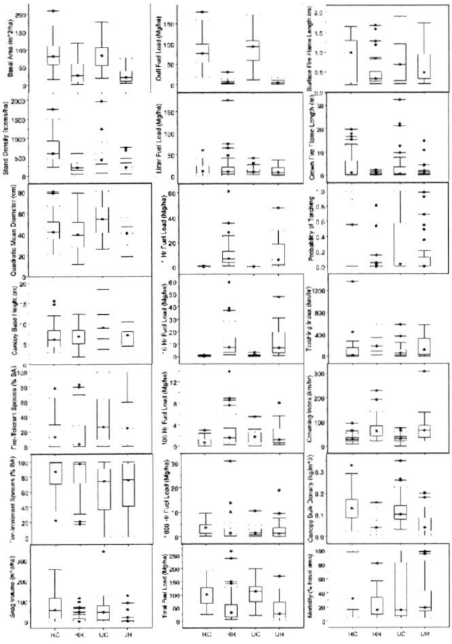

Figure 7. Box and whisker plots of current and reconstructed stand structure, fuel load, and fire behavior and effects variables for riparian and upland forests. RC is riparian current, RR is riparian reconstructed, UC is upland current, UR is upland reconstructed. Boxes are the upper and lower quartiles divided at the median, whiskers are the maximum and minimum values, dots are outliers.

Table 4. Comparison of riparian vs. upland and current vs. reconstructed stand structure, fuel loads, and potential fire behavior and effects least squares mean (standard error) values. BA is basal area, QMD is quadratic mean diameter, CBH is crown base height (average height to lowest green branch), CBD is canopy bulk density. Values in the same row followed by a different letter are significantly different (Tukey's post-hoc ANOVA, p<0.05), p-values are for the ANOVA global F test. Sample size is 36 for all variables in each column except riparian reconstructed (n=35 for torching index, crowning index, and CBD), and upland reconstructed (n=35 for all variables). Fire behavior calculated under 97th weather conditions.

	Riparian		Upland	
	Current	Reconstructed	Current	Reconstructed
BA (m^2/ha)	87.4(0.2)a	28.5(0.2)b	77.7(0.2)a	21.4(0.2)b
Stand Dens(stems/ha)	634.5(1.1)a	207.7(1.1)b	401.4(1.1)c	201.1(1.1)b
QMD (cm)	45.7(2.6)a	40.0(2.6)ab	55.3(2.6)c	38.4(2.6)b
Avg CBH (m)	6.7(0.01)a	6.5(0.01)a	9.4(0.01)b	6.3(0.01)a
% Comp (by BA)				
Fire Tolerant$_1$	13.4(0.5)ab	10.1(0.5)a	36.3(0.5)c	30.3(0.5)bc
Fire Intolerant$_2$	86.6(0.5)ab	89.9(0.5)a	63.7(0.5)c	69.7(0.5)bc
Snag Vol (m^3/ha)	36.8(0.4)a	2.4(0.4)b	24.0(0.4)a	0.5(0.4)b
Fuel Loads (Mg/ha)				
Duff	69.1(1.2)a	3.3(1.2)b	72.2(1.2)a	3.0(1.2)b
Litter	13.0(1.2)a	8.8(1.2)a	12.3(1.2)a	6.9(1.2)a
1 hr	0.1(1.2)a	5.4(1.2)b	0.1(1.2)a	3.0(1.2)b
10 hr	0.4(1.2)a	6.3(1.2)b	0.5(1.2)a	5.5(1.2)b
100 hr	0.7(0.01)a	1.8(0.01)b	1.4(0.01)ab	1.5(0.01)ab
1000 hr	2.8(0.0)a	1.4(0.0)ab	1.1(0.0)ab	0.9(0.0)b
Total	92.5(1.2)a	27.9(1.2)b	91.1(1.2)a	22.3(1.2)b
Flame Length (m)				
Surface	0.6(1.1)a	0.4(1.1)b	0.6(1.1)a	0.5(1.1)ab
Crown	0.9(1.1)a	0.4(1.1)b	0.6(1.1)ac	0.5(1.1)bc
Prob of Torch	0.45(0.06)a	0.03(0.06)b	0.22(0.06)c	0.08(0.06)bc

Torch Index (km/hr)	20.1(0.4)a	176.3(0.4)b	47.1(0.4)ac	98.6(0.4)bc
Crown Index (km/hr)	27.5(0.1)a	61.6(0.1)b	28.8(0.1)a	61.9(0.1)b
CBD (kg/m³)	0.12(1.14)a	0.04(1.14)b	0.10(1.14)a	0.04(1.14)b
Mortality (% BA)	30.6(1.2)a	16.5(1.2)b	15.7(1.2)b	21.0(1.2)ab

[1]Pinus jeffreyi, P. ponderosa, P. lambertiana, P. monticola, and Quercuz kelloggii.

[2]P. contorta ssp. murrayana, Populus tremuloides, P. balsamifera ssp. trichocarpa, Alnus incana ssp. tenuifolia, Salix spp., Abies concolor, A. magnifica, Calocedrus decurrens, and Pseudotsuga menziesii.

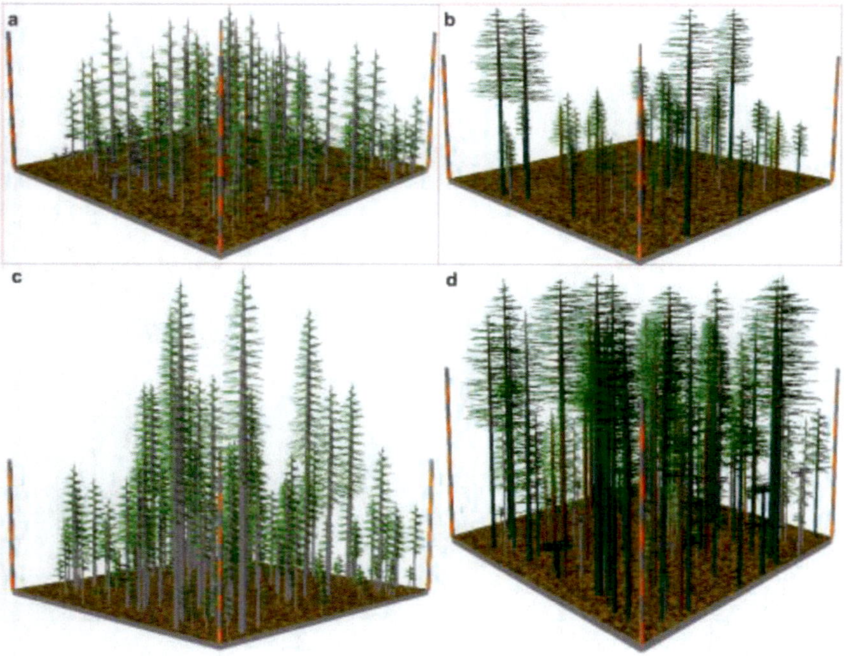

Figure 8. Stand visualization simulation of typical conditions for a) current riparian forest (Dollar Creek, 2009), and b) reconstructed riparian forest (West Branch Feather River, 1886). The corresponding stands, c) Dollar Creek riparian, reconstructed conditions in 1962, and d) West Branch Feather River, current conditions in 2009, are not representative of typical conditions but are displayed for comparison. Stands representative of typical conditions (outlined in red) were selected based on how close the stand density, basal area, and species composition values of the individual stands were to the mean values for all sites. Range pole intervals are approximately 3 m, ground area is approximately 0.75 ha.

Current riparian stands have more than triple the BA and stem density (Figure 8), more than 15 times the snag volume, more than 20 times the duff load, and more than triple the total fuel load of reconstructed riparian stands. However, woody fuel loads are much lower in current riparian forests in all size classes except the 1000 hr fuels. Potential flame lengths in current riparian stands are 50% greater than those of reconstructed conditions for surface fires, and 125% greater for crown fires.

The probability of torching has increased by a factor of 15, the torching index is an order of magnitude less, and the crowning index (i.e., the wind speed required to initiate active crown fire) is less than half that of reconstructed riparian stands. Current riparian stands have triple the CBD, and nearly double the predicted mortality of reconstructed riparian conditions.

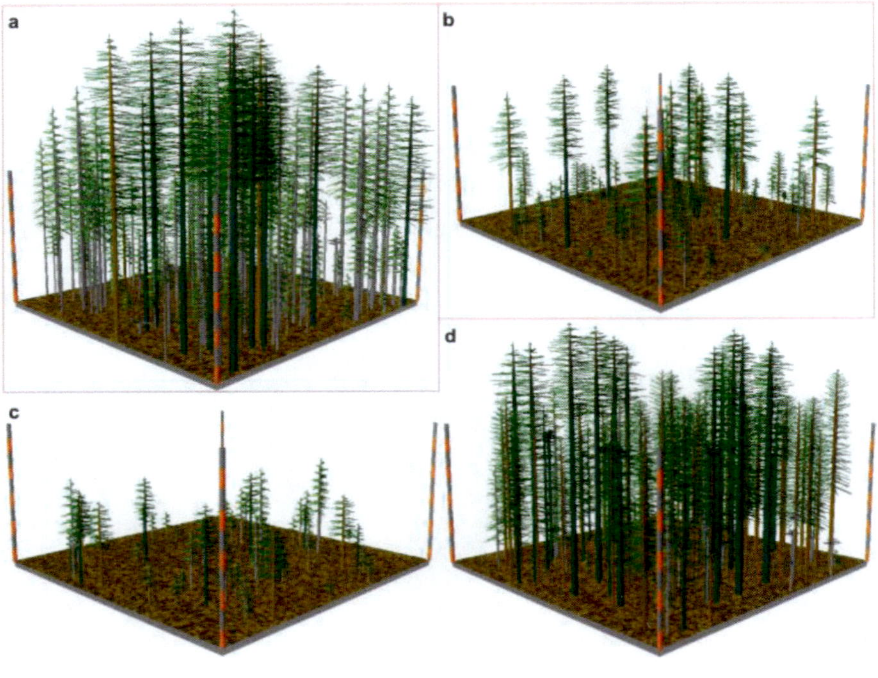

Figure 9. Stand visualization simulation of typical conditions for a) current upland forest (Jones Creek, 2009), and b) reconstructed upland forest (B Fork Onion Creek, 1904). The corresponding stands, c) Jones Creek upland, reconstructed conditions in 1869, and d) B Fork Onion Creek, current conditions in 2009, are not representative of typical conditions but are displayed for comparison. Stands representative of typical conditions (outlined in red) were selected based on how close the stand density, basal area, and species composition values of the individual stands were to the mean values for all sites. Range pole intervals are approximately 3 m, ground area is approximately 0.75 ha.

Current and reconstructed upland stands were significantly different in BA, stand density, snag volume, QMD, average CBH, duff, 1 hr, 10 hr, and total fuel load, crowning index, and CBD (Table 4). Reconstructed upland stands have less than one third the BA, half the stem density, and two orders of magnitude lower snag volume of current upland stands (Figure 9). The QMD of current upland forests has increased by nearly 50%, while the CBH has increased by three meters. Duff load is over 24 times greater and total fuel load is double in current upland forests, while 1 hr and 10 hr woody fuel loads are lower by an order of magnitude. Crowning index under reconstructed upland forest conditions is more than double that of current conditions, while CBD is less than half.

Current riparian and upland forests had significantly different stand density, QMD, average CBH, species composition, probability of torching, and predicted mortality (Figure 8). Current riparian forests have greater than 50% more stems per ha, a QMD nearly 20% lower, and a CBH that is nearly 3 m lower. In current riparian stands, fire-tolerant species comprise nearly 16 % less of the total basal area than current upland stands. The probability of torching in current riparian stands is more than double that of current upland stands, and the predicted mortality is nearly double.

Reconstructed riparian forests were not significantly different from reconstructed upland forests in any of the variables analyzed. Site-level data is provided here as a supplement.

SUPPLEMENTAL DATA

Site-level data for a) current riparian, b) reconstructed riparian, c) current upland, and d) reconstructed upland stands. BA is basal area, QMD is quadratic mean diameter, CBH is crown base height, % Comp is species composition by basal area lumped into fire-tolerant and fire-intolerant groups, Prob Torch is the probability of torching, CBD is canopy bulk density, C1 FRI is a broad filter fire return interval (one or more trees scarred per site), C10 FRI is a narrow filter fire return interval (two or more trees scarred per site), Recon Year is the year of reconstruction (year of the last fire), Fuel Accum Time is the fuel accumulation time (period between the last fire and the second to last fire).

Supplemental Data

a

Area	Site	BA	Stand Dens	Snag vol	QMD	Avg CBH	% Comp (by BA)		Canopy	Seedlings	Shrub	Fuel Load (Mg/ha)							Flame Length (m)		Prob	Torch Index	Crown Index	CBD	Mortality
		(m^2/ha)	(stems/ha)	(m^3/ha)	(cm)	(m)	Fire Tol	Fire Intol	Cover (%)	(stems/ha)	Cover (%)	Duff	Litter	1 hr	10 hr	100 hr	1000 hr	Total	Surface	Crown	Torch	(km/hr)	(km/hr)	(kg/m^3)	(% BA)
Lassen National Forest	BT	138	880	9	51	9.4	30	70	64	0	1	34	11	0.1	0.4	0.9	0.3	46	0.2	18.0	0.00	251	18	0.23	100
	CA	147	950	68	47	11.8	9	91	83	0	2	70	13	0.1	0.8	0.9	5.2	90	0.2	19.8	0.00	445	15	0.29	100
	EL	149	610	3	67	8.2	34	67	47	2000	8	104	29	0.1	0.4	0.4	4.3	138	0.2	0.3	0.00	209	30	0.12	6
	FI	62	520	165	40	8.0	12	88	70	3000	4	83	11	0.2	0.4	1.5	4.1	100	1.3	1.2	0.54	27	28	0.13	32
	FR	145	340	116	79	14.7	66	34	47	3000	16	55	43	0.1	1.1	1.5	0.7	102	1.0	0.9	0.00	156	41	0.07	5
	JO	125	380	69	67	12.1	25	76	65	12000	1	87	26	0.1	0.2	0.4	5.4	119	0.7	0.6	0.00	111	42	0.07	16
	LC	79	620	115	42	10.2	13	87	67	3000	2	91	27	0.3	0.8	1.5	1.8	122	1.4	13.4	0.79	0	22	0.17	99
	PB	33	470	9	33	4.1	33	68	60	0	65	55	2	0.1	0.3	0.2	0.1	58	0.3	0.3	0.17	155	54	0.05	15
	RO	77	450	60	50	4.7	0	100	48	4000	16	74	9	0.1	0.3	0.2	2.0	86	1.4	6.1	0.75	7	28	0.13	96
	SH	101	520	56	55	5.3	29	71	45	15000	0	66	39	0.1	0.4	0.4	1.2	107	1.2	4.9	0.93	8	32	0.11	93
	SL	48	590	11	37	2.8	0	100	47	2000	12	53	3	0.1	0.3	0.0	6.6	63	1.3	6.1	0.90	2	31	0.11	98
	SM	76	760	102	38	4.2	0	100	78	1000	4	57	7	0.2	0.4	1.1	1.1	67	1.3	9.4	0.99	0	26	0.14	98
	SN	100	920	42	37	9.5	0	100	45	2000	25	70	18	0.4	0.7	0.7	3.8	94	1.2	18.0	0.67	11	20	0.20	99
	WF	144	1030	152	43	8.2	0	100	61	2000	1	136	7	0.1	0.5	0.2	0.5	143	1.2	1.2	0.83	21	30	0.12	18
	WL	47	960	23	26	4.1	1	99	48	1000	1	42	22	0.1	0.3	0.9	3.5	69	1.3	10.4	0.94	7	23	0.17	99
Onion Creek	OB	96	380	40	53	6.3	20	80	38	4000	58	17	5	0.1	0.4	1.7	7.3	32	0.2	0.3	0.00	135	20	0.18	13
	OC	209	420	48	81	15.6	4	96	61	15000	35	87	37	0.3	0.5	1.5	4.1	130	0.2	0.3	0.00	276	33	0.09	5
	OD	91	500	60	50	6.8	8	93	43	2000	51	87	13	0.1	0.6	0.9	4.6	106	0.2	0.3	0.00	130	40	0.07	10
	OG	169	840	114	54	8.6	3	97	70	0	12	119	18	0.6	0.3	0.7	0.8	139	1.6	16.8	0.82	0	17	0.16	99
Lake Tahoe Basin, West Side	BC	95	1465	29	44	2.9	3	97	100	400	37	159	7	0.1	0.4	1.3	1.1	169	0.6	0.9	0.95	0	38	0.08	37
	BW	83	910	139	30	4.9	38	62	55	1000	38	19	4	0.0	0.1	0.2	1.7	25	0.2	0.3	0.00	155	21	0.16	23
	DC	100	750	257	57	5.9	2	98	28	1000	15	74	6	0.1	0.3	0.7	6.6	88	0.2	0.3	0.00	194	24	0.15	15
	GC	58	1520	210	21	2.8	43	57	43	2000	27	106	6	0.1	0.5	0.4	3.4	116	1.0	4.6	1.00	1	15	0.28	98
	ME	15	240	21	29	4.0	0	100	46	1000	53	36	8	0.0	0.4	0.7	5.6	50	1.2	1.5	0.57	4	49	0.06	94
	MK	104	1010	239	62	5.3	27	73	68	900	30	95	14	0.1	0.5	2.9	2.1	115	0.2	0.3	0.00	239	13	0.33	14
	RC	56	410	161	46	5.9	21	79	44	1000	62	68	8	0.1	0.3	0.4	0.6	77	1.3	2.4	0.71	16	34	0.10	78
	RU	50	400	0	42	6.3	0	100	6	0	15	19	4	0.0	0.1	0.0	5.2	29	1.5	6.7	0.65	6	27	0.13	98
	TC	63	520	11	38	4.9	0	100	33	5000	20	121	61	0.2	0.5	0.2	9.7	192	0.9	0.9	0.87	10	23	0.16	26
	TL	39	650	5	40	3.8	0	100	50	11000	2	32	0	0.0	0.0	0.0	0.9	33	0.2	0.3	0.00	1360	94	0.02	38
	WD	36	480	20	36	6.8	49	51	43	1000	38	83	20	0.1	0.3	0.7	5.3	109	1.0	0.9	0.25	42	28	0.12	30
Lake Tahoe Basin, East Side	BU	18	280	22	29	3.4	78	22	21	1000	23	26	42	0.0	0.1	0.2	4.3	73	0.6	0.6	0.18	21	63	0.04	17
	HT	66	490	55	42	10.2	27	74	74	4000	66	78	18	0.1	0.2	1.5	2.7	101	1.0	1.2	0.42	8	31	0.10	26
	MA	64	530	121	41	6.7	23	77	53	2000	14	119	12	0.1	0.6	2.4	4.5	138	1.1	0.9	0.37	23	21	0.17	26
	MF	98	1350	108	31	6.6	3	97	59	1000	3	85	34	0.1	0.7	1.3	4.0	125	0.9	0.9	0.92	10	19	0.19	32
	TU	73	1760	0	27	4.8	33	67	60	9000	100	150	9	0.1	0.4	0.7	1.0	161	1.6	15.5	0.99	0	8	0.29	100
	ZC	117	1220	3	43	4.6	24	76	80	2000	38	178	5	0.0	0.3	0.0	3.8	187	1.0	1.5	0.90	0	65	0.04	30

b

Area	Site	C1 FRI (yrs)	C10 FRI (yrs)	Recon Year	Fuel Accum Time (yrs)	BA (m^2/ha)	Stand Dens (stems/ha)	Snag vol (m^3/ha)	QMD (cm)	Avg CBH (m)	% Comp (by BA) Fire Tol	Fire Intol	Fuel Loading (Mg/ha) Duff	Litter	1 hr	10 hr	100 hr	1000 hr	Total	Flame Length (m) Surface	Crown	Prob Torch	Torch Index (km/hr)	Crown Index (km/hr)	CBD (kg/m^3)	Mortality (% BA)
Lassen National Forest	BT	19.2	24.3	1976	7	84	500	10	52	8.3	38	62	7	9	6.5	4.4	1.3	0.9	29	0.3	0.3	0.00	268	35	0.10	9
	CA	9.5	--	1862	11	9	150	0	27	5.0	15	85	1	5	2.3	2.3	0.4	0.4	12	0.5	0.3	0.00	185	92	0.03	34
	EL	21.4	61.3	1980	96	113	270	0	79	12.4	37	63	30	75	61.0	59.4	8.5	30.9	265	0.3	0.3	0.00	336	66	0.04	4
	FI	12.5	15.3	1900	20	10	260	0	23	4.0	8	92	3	4	4.9	4.9	0.7	0.8	19	0.4	0.3	0.00	147	81	0.03	29
	FR	14.5	--	1886	35	39	230	0	46	7.8	69	31	15	22	6.6	7.1	3.3	1.2	56	1.2	1.2	0.01	89	61	0.04	13
	JO	17.7	44.5	1958	77	99	440	0	54	7.7	49	51	14	66	10.1	28.4	8.9	3.2	130	0.7	0.6	0.00	94	42	0.08	8
	LC	16.0	--	1858	18	10	50	0	56	7.4	0	100	2	7	7.9	7.9	0.9	4.2	30	0.5	0.6	0.00	186	228	0.01	4
	PB	13.0	10.0	1944	15	1	120	0	12	1.8	0	100	3	2	1.7	1.9	0.2	0.1	8	0.5	0.6	0.03	102	144	0.01	57
	RO	21.0	23.0	1974	7	46	250	70	50	8.9	0	100	1	3	3.5	3.6	0.7	1.1	13	0.3	0.3	0.00	186	49	0.06	12
	SH	15.7	24.3	1903	29	28	210	0	40	7.1	27	73	7	13	10.6	10.8	2.0	2.9	47	0.4	0.3	0.00	186	63	0.04	13
	SL	20.5	20.5	1957	20	20	310	0	33	6.0	0	100	2	11	3.2	2.8	0.7	0.6	20	1.7	1.5	0.81	21	57	0.05	56
	SM	11.9	17.9	1973	27	37	340	113	39	6.7	0	100	4	9	11.1	11.2	1.7	2.8	40	1.6	2.4	0.54	14	39	0.08	82
	SN	15.9	17.7	1911	75	5	200	0	17	3.0	0	100	11	13	16.2	16.1	1.8	2.0	59	0.4	0.3	0.00	96	67	0.04	42
	WF	25.7	--	1929	63	83	220	0	68	10.6	0	100	10	49	22.9	36.7	13.9	9.9	142	0.3	0.3	0.00	442	58	0.05	7
	WL	8.5	17.1	1918	56	5	60	0	33	6.3	0	100	8	20	25.3	25.1	2.8	5.7	87	0.5	0.6	0.00	338	95	0.02	23
Onion Creek	OB	22.2	35.7	1904	22	25	140	0	48	6.7	3	97	4	12	4.0	7.2	3.4	1.9	32	0.3	0.3	0.00	586	61	0.04	6
	OC	15.2	36.5	1903	31	74	340	96	52	9.5	3	97	5	16	12.1	14.4	4.4	3.3	55	0.2	0.3	0.00	409	37	0.08	12
	OD	20.3	32.0	1878	5	25	170	0	41	5.8	0	100	1	2	0.8	1.3	0.5	0.2	5	0.3	0.3	0.00	130	84	0.03	13
	OG	12.0	25.8	1886	15	31	240	0	41	6.8	0	100	2	6	6.8	7.1	1.1	2.2	26	0.4	0.3	0.00	338	54	0.04	13
Lake Tahoe Basin – West Side	BC	20.2	22.5	1923	24	56	200	0	60	8.6	2	98	4	16	6.2	11.3	4.9	2.3	45	0.2	0.3	0.00	425	49	0.06	6
	BW	42.3	--	1952	16	42	430	48	38	6.9	18	82	2	6	5.9	6.3	0.8	1.5	22	0.2	0.3	0.00	290	22	0.16	16
	DC	14.4	20.8	1962	23	118	600	0	53	8.8	1	99	3	11	13.4	13.4	1.5	5.3	47	0.2	0.3	0.00	197	27	0.13	10
	GC	16.0	56.0	1963	94	72	400	20	54	8.6	16	85	13	42	35.3	38.5	7.6	13.7	150	0.2	0.3	0.00	212	40	0.07	9
	ME	15.6	35.5	1848	41	24	210	0	39	7.9	0	100	4	48	6.9	4.1	1.6	1.2	66	1.3	1.2	0.15	45	62	0.04	47
	MK	10.0	32.0	1945	53	98	410	0	71	12.2	30	71	7	27	22.2	23.7	3.1	0.4	84	0.2	0.3	0.00	208	37	0.08	6
	RC	19.6	86.5	1956	5	61	450	0	42	7.9	7	93	1	2	2.6	2.6	0.3	0.8	9	0.3	0.3	0.00	499	28	0.13	16
	RU	14.8	--	1938	96	10	180	0	27	5.4	0	100	13	40	28.1	26.2	3.2	4.9	115	0.3	0.3	0.00	171	62	0.04	40
	TC	8.4	23.0	1860	7	3	50	0	25	2.4	0	100	1	2	1.4	1.7	0.3	0.2	7	0.3	0.3	0.00	--	--	--	35
	TL	16.1	--	1977	112	8	50	4	46	10.4	0	100	12	175	25.0	15.6	6.0	4.5	238	0.3	0.3	0.00	506	105	0.02	36
	WD	22.8	39.0	1912	48	3	100	65	21	3.9	79	21	6	10	4.0	6.8	1.4	0.4	28	0.8	0.9	0.00	62	191	0.01	33
Lake Tahoe Basin – East Side	BU	14.3	26.0	1959	9	17	260	35	29	5.2	24	76	1	3	2.4	2.7	0.4	0.5	10	0.6	0.6	0.00	56	44	0.07	22
	HT	9.0	--	1937	36	28	430	45	29	6.4	25	76	5	10	7.8	9.5	1.9	1.4	36	0.2	0.3	0.00	227	64	0.04	27
	MA	22.1	32.5	1863	8	14	220	0	27	4.5	57	44	1	2	1.5	1.9	0.3	0.3	7	0.3	0.3	0.00	168	75	0.03	24
	MF	10.7	17.0	1919	16	26	240	44	38	7.0	1	99	2	12	4.6	4.1	0.7	1.1	24	0.3	0.3	0.00	195	66	0.04	28
	TU	16.8	29.6	1924	49	7	160	0	24	3.7	83	17	7	10	5.2	8.0	1.6	0.5	32	0.4	0.6	0.00	87	89	0.02	19
	ZC	12.0	14.6	1864	7	5	80	0	29	3.9	40	60	1	2	0.7	1.3	0.4	0.1	5	0.9	0.9	0.05	65	141	0.01	16

c

Area	Site	BA (m²/ha)	Stand Dens (stems/ha)	Snag vol (m³/ha)	QMD (cm)	Avg CBH (m)	% Comp (by BA) Fire Tol	% Comp (by BA) Fire Intol	Canopy Cover (%)	Seedlings (stems/ha)	Shrub Cover (%)	Fuel Load (Mg/ha) Duff	Litter	1 hr	10 hr	100 hr	1000 hr	Total	Flame Length (m) Surface	Crown	Prob Torch	Torch Index (km/hr)	Crown Index (km/hr)	CBD (kg/m³)	Mortality (% BA)
Lassen National Forest	BT	85	410	0	54	6.8	30	70	45	100	2	106	11	0.2	1.1	2.6	2.6	123	0	0	0.00	248	33	0.11	11
	CA	178	380	0	78	15.9	24	76	69	1000	1	68	8	0.2	0.6	1.3	4.6	83	0	0	0.00	365	28	0.13	5
	EL	69	340	81	53	5.9	26	74	68	5000	8	104	10	0.1	0.3	0.7	0.6	115	0	0	0.00	132	44	0.07	9
	FI	92	750	47	41	18.4	14	86	83	0	0	95	16	0.3	0.7	2.0	4.6	119	1	22	0.00	37	15	0.27	100
	FR	156	350	48	76	15.1	25	75	62	4000	36	34	9	0.1	0.2	1.2	0.3	44	1	13	0.64	12	22	0.10	95
	JO	113	540	43	56	13.6	35	65	50	2000	1	110	15	0.1	0.5	2.4	2.1	130	1	1	0.00	73	39	0.08	9
	LC	71	750	93	36	8.1	50	50	73	0	0	157	17	0.2	1.0	2.8	1.0	179	1	2	0.74	16	32	0.11	69
	PB	69	620	4	39	13.0	9	91	83	0	0	172	18	0.3	3.3	5.4	0.2	199	0	0	0.00	582	32	0.11	17
	RO	100	450	34	56	8.4	0	100	42	3000	15	106	17	0.1	0.6	2.0	0.9	126	1	4	0.57	7	37	0.09	95
	SH	96	440	17	55	7.7	23	77	40	16000	17	104	8	0.2	1.5	2.8	3.6	120	2	21	0.74	0	19	0.18	100
	SL	104	630	14	50	9.7	0	100	54	1000	0	51	4	0.1	0.1	0.7	1.7	58	1	8	0.70	0	30	0.11	97
	SM	39	500	105	37	5.6	63	37	39	0	3	119	10	0.1	0.7	0.4	1.7	131	1	3	0.81	16	29	0.12	72
	SN	107	900	56	41	9.3	3	97	80	2000	5	89	18	0.1	0.5	2.0	1.0	110	1	22	0.55	8	16	0.26	100
	WF	138	430	47	68	14.0	10	90	44	0	10	114	9	0.2	1.0	5.5	1.3	131	1	7	0.57	0	37	0.09	94
	WL	40	280	1	49	5.5	85	16	55	1000	0	127	11	0.1	0.6	1.1	10.5	150	1	1	0.39	28	61	0.05	16
Onion Creek	OB	113	310	47	74	12.1	40	60	23	12000	0	78	42	0.1	0.4	2.4	4.4	128	0	0	0.00	155	35	0.08	7
	OC	116	470	71	64	12.0	6	94	42	13000	25	84	7	0.2	0.5	1.5	0.6	94	0	0	0.06	59	28	0.09	5
	OD	110	320	126	51	16.4	1	99	72	0	0	121	30	0.2	1.5	0.4	3.1	156	0	0	0.00	234	25	0.14	12
	OG	79	430	93	50	7.6	0	100	73	0	0	102	31	0.1	0.7	2.0	2.4	138	1	1	0.49	16	31	0.10	19
Lake Tahoe Basin – West Side	BC	89	870	118	41	5.7	38	63	72	200	2	92	3	0.1	0.0	0.5	2.3	98	2	32	1.00	0	0	0.35	100
	BW	107	345	0	66	8.6	63	37	70	0	19	96	15	0.1	0.6	2.5	0.1	114	0	0	0.00	108	55	0.05	7
	DC	19	80	62	63	11.0	58	42	25	0	21	58	12	0.1	0.5	3.1	1.2	75	1	1	0.00	120	45	0.06	8
	GC	48	455	75	51	5.8	22	78	45	1320	14	112	13	0.0	0.3	1.9	3.8	131	1	2	0.38	5	36	0.08	42
	ME	71	190	36	81	8.3	10	91	10	0	16	134	16	0.2	0.8	1.9	3.1	156	0	0	0.00	190	43	0.06	6
	MK	104	451	342	78	6.7	64	36	50	2700	7	95	7	0.0	0.2	0.7	0.0	103	0	0	0.60	40	17	0.23	6
	RC	76	675	120	59	10.3	12	88	70	60	2	38	6	0.1	0.8	2.6	1.4	49	1	1	0.46	19	21	0.17	17
	RU	71	270	0	55	3.7	0	100	10	0	18	20	4	0.0	0.2	1.0	0.8	27	0	0	0.00	96	26	0.12	18
	TC	35	115	1	73	13.3	86	14	10	120	9	27	5	0.0	0.1	0.0	0.0	32	1	1	0.18	23	74	0.03	6
	TL	62	1960	101	26	5.0	23	77	75	0	4	98	15	0.1	1.2	1.7	2.9	119	1	10	0.98	2	11	0.35	99
	WD	162	600	34	61	10.0	83	18	70	360	23	46	14	0.0	0.3	0.3	0.0	61	1	1	0.00	79	16	0.23	10
Lake Tahoe Basin – East Side	BU	25	180	0	42	9.3	92	8	35	0	9	36	10	0.0	0.4	0.5	0.8	47	1	1	0.00	77	65	0.04	10
	HT	48	230	0	66	12.6	100	0	20	60	5	62	8	0.0	0.3	0.7	0.6	72	1	1	0.00	183	52	0.05	6
	MA	97	315	52	76	10.5	4	97	40	300	5	61	27	0.1	0.4	3.1	4.7	96	0	0	0.00	234	32	0.10	6
	MF	33	410	41	38	5.9	84	17	45	0	40	84	11	0.0	0.5	3.0	1.9	100	1	1	0.00	98	28	0.11	15
	TU	58	390	10	46	7.1	100	0	41	0	17	74	6	0.0	0.6	1.1	1.3	83	1	1	0.00	102	38	0.08	8
	ZC	42	1240	21	31	4.8	91	9	35	0	31	12	9	0.0	0.0	0.0	0.2	21	1	1	0.25	51	19	0.15	34

d

Area	Site	C1 FRI (yrs)	C10 FRI (yrs)	Recon Year	Fuel Accum Time (yrs)	BA (m^2/ha)	Stand Dens (stems/ha)	Snag vol (m^3/ha)	QMD (cm)	Avg CBH (m)	% Comp (by BA) Fire Tol	% Comp (by BA) Fire Intol	Fuel Loading (Mg/ha) Duff	Fuel Loading (Mg/ha) Litter	Fuel Loading (Mg/ha) 1 hr	Fuel Loading (Mg/ha) 10 hr	Fuel Loading (Mg/ha) 100 hr	Fuel Loading (Mg/ha) 1000 hr	Fuel Loading (Mg/ha) Total	Flame Length (m) Surface	Flame Length (m) Crown	Prob Torch	Torch Index (km/hr)	Crown Index (km/hr)	CBD (kg/m^3)	Mortality (% BA)
Lassen National Forest	BT	8.8	19.1	1914	23	16	170	0	34	5.9	52	48	3	8	6.5	7.7	1.2	1.2	28	0	0	0.00	21	74	0.03	16
	CA	13.8	24.7	1870	19	23	200	0	38	7.2	41	59	5	8	5.7	5.8	1.2	1.1	27	0	0	0.00	133	64	0.04	17
	EL	16.5	27.3	1931	28	34	240	0	40	7.5	24	76	6	13	13.9	13.6	1.8	3.6	52	0	0	0.00	336	49	0.06	16
	FI	12.3	28.0	1956	73	25	740	0	23	4.0	10	90	13	18	18.3	17.9	2.4	2.6	72	0	0	0.00	170	19	0.20	34
	FR	13.9	18.2	1928	42	68	270	0	53	9.2	21	79	10	25	20.7	22.1	4.1	9.4	92	1	1	0.00	319	40	0.06	8
	JO	13.0	12.2	1869	6	9	190	0	26	4.8	47	53	1	2	1.3	1.3	0.2	0.2	6	1	1	0.17	30	82	0.03	37
	LC	12.0	--	1910	26	6	200	0	20	3.6	59	41	3	5	3.0	4.1	0.7	0.4	16	1	1	0.49	18	83	0.03	86
	PB	42.5	42.5	1929	75	6	230	0	19	3.4	0	100	11	15	16.6	17.1	2.5	2.2	64	0	1	0.00	101	83	0.03	54
	RO	14.8	35.3	1955	50	47	270	0	47	9.2	0	100	7	23	24.2	25.6	4.8	7.3	92	0	0	0.00	204	38	0.08	16
	SH	58.0	52.0	1973	70	52	420	20	41	7.3	31	69	15	29	26.8	26.8	4.5	6.1	109	2	4	0.35	0	42	0.07	95
	SL	11.8	--	1975	9	59	470	2	43	7.9	0	100	1	3	3.0	3.3	0.7	0.8	12	2	5	0.55	11	31	0.11	98
	SM	13.5	46.0	1965	69	32	680	0	31	4.5	6	94	10	20	17.6	19.2	2.0	2.8	72	1	11	0.92	9	22	0.18	98
	SN	23.0	46.0	1980	29	45	320	58	34	6.3	2	98	4	9	11.7	11.8	1.4	2.8	41	1	15	0.69	16	20	0.20	99
	WF	31.8	32.7	1929	63	49	350	0	44	7.7	22	79	9	27	27.5	29.5	4.8	7.6	106	1	2	0.12	20	34	0.10	34
	WL	10.8	14.6	1967	43	19	220	0	40	5.5	93	7	6	12	3.3	6.8	1.8	0.6	31	1	1	0.21	51	99	0.02	16
Onion Creek	OB	15.8	21.4	1904	17	24	240	0	41	7.0	63	37	3	6	3.3	4.2	0.6	0.0	17	1	1	0.02	61	48	0.06	15
	OC	13.4	20.3	1872	10	19	130	0	42	7.8	0	100	1	4	4.6	4.8	0.9	1.2	17	0	0	0.00	395	46	0.05	17
	OD	24.8	42.6	1960	87	75	400	0	48	7.9	0	100	13	38	47.7	47.6	5.6	18.8	170	0	0	0.00	267	31	0.10	13
	OG	7.4	15.6	1886	13	10	210	0	23	3.5	0	100	2	3	3.2	3.3	0.6	0.7	13	0	0	0.00	117	82	0.03	28
Lake Tahoe Basin, West Side	BC	13.8	23.5	1890	4	27	155	0	53	8.3	4	97	1	2	1.1	1.5	0.5	0.4	6	0	0	0.00	400	40	0.06	8
	BW	19.4	13.4	1892	16	12	235	0	26	5.0	60	41	2	5	2.0	3.1	0.6	0.4	13	0	0	0.00	216	76	0.03	35
	DC	13.7	26.5	1911	9	9	45	0	56	8.7	68	32	1	4	1.4	2.6	0.6	0.7	10	1	1	0.00	153	135	0.01	4
	GC	17.4	54.6	1990	87	32	355	91	47	8.3	27	73	13	29	28.2	30.7	4.1	5.9	110	0	0	0.00	354	27	0.11	19
	ME	11.7	10.0	1888	1	20	95	0	45	6.5	4	96	0	1	0.0	0.3	0.2	0.0	1	0	0	0.00	373	85	0.03	9
	MK	11.4	56.3	1892	30	45	250	0	50	9.0	41	59	4	13	6.2	9.7	3.2	1.8	38	0	0	0.00	405	36	0.09	10
	RC	17.6	37.5	1957	67	38	225	92	55	10.4	37	63	17	36	29.7	29.9	4.9	7.3	124	0	0	0.00	426	43	0.07	11
	RU	29.3	--	1930	60	28	120	0	42	8.0	0	100	9	27	22.5	25.9	8.0	5.7	98	0	0	0.00	281	114	0.02	19
	TC	15.1	32.2	1883	12	5	55	0	32	4.4	100	0	1	3	0.1	1.4	0.5	0.0	6	1	1	0.03	49	307	0.00	14
	TL	11.6	--	1980	74	20	760	126	30	3.6	7	93	10	11	10.5	11.8	1.6	1.9	47	1	2	0.98	1	25	0.14	67
	WD	18.7	29.0	1882	18	10	360	0	20	3.4	56	44	2	3	0.6	1.9	0.5	0.1	9	1	1	0.05	39	101	0.02	43
Lake Tahoe Basin, East Side	BU	6.1	11.1	1896	--	--	--	--	--	--	--	--	--	--	--	--	--	--	--	--	--	--	--	--	--	--
	HT	10.7	--	1957	38	18	230	0	45	7.2	100	0	5	11	0.3	4.9	1.7	0.1	22	1	1	0.02	58	79	0.03	44
	MA	16.9	21.3	1867	9	13	165	0	33	4.4	3	97	2	3	0.6	1.7	1.1	0.3	9	0	0	0.00	155	92	0.02	14
	MF	10.9	16.8	1917	14	14	70	0	49	9.0	3	97	2	8	9.2	9.3	1.1	3.1	32	0	0	0.00	575	139	0.01	10
	TU	11.6	16.5	1932	7	9	220	0	24	4.1	100	0	1	2	0.0	0.7	0.2	0.0	3	1	1	0.04	49	99	0.02	29
	ZC	13.4	14.2	1935	7	10	100	0	43	7.1	97	3	1	2	1.0	1.7	0.4	0.1	6	1	1	0.00	46	75	0.03	20

3.6. Correlation of Reconstructed Variables with Fire Return Interval

Table 5. Pearson's correlation coefficients exploring the relationships between a broad filter fire return interval (C1 FRI, derived from all fire events scarring one or more trees at a given site) and reconstructed riparian and upland stand structure, fuel loads, and potential fire behavior and effects. * indicates significant correlation (p<0.05). Sample size is 36 for all variables in each column except riparian (n=35 for torching index, crowning index, and CBD), and upland (n=35 for all variables)

	Riparian	Upland
BA (m^2/ha)	0.207	0.289
Stand Density (stems/ha)	0.190	0.111
QMD (cm)	0.223	-0.035
Avg CBH (m)	0.190	0.040
% Composition (by BA)		
Fire Tolerant	0.133	-0.170
Fire Intolerant	-0.133	0.170
Snag Volume (m^3/ha)	0.094	0.046
Fuel Loads (Mg/ha)		
Duff	0.062	0.524*
Litter	0.044	0.488*
1 hr	0.013	0.507*
10 hr	0.076	0.496*
100 hr	0.206	0.501*
1000 hr	0.140	0.380*
Total	0.070	0.520*
Flame Length (m)		
Surface	-0.094	0.311
Crown	-0.112	0.12
Probability of Torching	0.003	0.037
Torching Index (km/hr)	0.236	-0.153
Crowning Index (km/hr)	-0.150	-0.100
CBD (kg/m^3)	0.415*	0.065
Mortality (% BA)	-0.252	0.331

Reconstructed riparian CBD was the only variable significantly correlated with C1 FRI in reconstructed riparian stands (Table 5). All reconstructed upland fuel-load variables (duff, litter, 1 hr, 10 hr, 100 hr, 1000 hr, total) were significantly correlated with upland C1 FRI. No other reconstructed upland variables were significantly correlated with C1 FRI in reconstructed upland stands.

3.7. Correlation of Current and Reconstructed Riparian and Upland Variables

Current riparian and upland stands have significantly correlated BA, snag volume, CBH, surface fire flame length, and probability of torching (Table 6). There were significant correlations between reconstructed riparian and upland QMD, average CBH, species composition, surface and crown fire flame length, probability of torching, and potential mortality. No other variables were significantly correlated between current or reconstructed riparian and upland stands.

Table 6. Pearson's analysis exploring the relationships between riparian and upland stand structure, fuel loads, and potential fire behavior and effects variables for both current and reconstructed conditions. * indicates significant correlation ($p<0.05$). Sample size is 36 for all variables in each column except reconstructed (n=35 for all variables except n=34 torching index, crowning index, and CBD)

	Current	Reconstructed
BA (m^2/ha)	0.384*	0.139
Stand Density (stems/ha)	0.172	-0.096
QMD (cm)	0.319	0.417*
Avg CBH (m)	0.485*	0.467*
% Composition (by BA)		
Fire Tolerant	0.322	0.355*
Fire Intolerant	0.322	0.355*
Snag Volume (m^3/ha)	0.511*	-0.107

Table 6. (Continued)

Fuel Loads (Mg/ha)		
Duff	-0.098	0.153
Litter	-0.207	0.027
1 hr	0.147	0.176
10 hr	-0.036	0.190
100 hr	-0.036	0.177
1000 hr	0.088	0.101
Total	-0.215	0.137
Flame Length (m)		
Surface	0.527*	0.366*
Crown	0.024	0.424*
Probability of Torching	0.418*	0.495*
Torching Index (km/hr)	0.104	0.161
Crowning Index (km/hr)	-0.147	0.219
CBD (kg/m^3)	-0.093	-0.003
Mortality (% BA)	0.229	0.494*

3.8. Comparison of Sampling Areas

While there was a great deal of variability within sampling areas, some interesting patterns emerge in the differences between them (Table 7). The east side of the Tahoe Basin consistently had the least fire-prone forest structure and fuel loads, while the Lassen National Forest was usually the most fire prone. East Tahoe had 55% the BA of Lassen, and 45% the BA of Onion Creek. Fire-tolerant species comprised 38.4% more of the BA in east Tahoe than in Onion Creek. Lassen had nearly double the duff fuel loads of east Tahoe. Fuel loads in the 1 hr size class in east Tahoe were 40% lower than Onion Creek, nearly 50% lower than west Tahoe, and more than 80% lower than Lassen. Fuel loads in the 10 and 1000 hr classes in east Tahoe were both more than 50% less than those of Lassen. Surface fire flame lengths in Lassen were nearly double those in Onion Creek and west Tahoe, while crown fire flame lengths in Lassen were nearly half those in Onion Creek.

Table 7. Comparison of stand structure, fuel loads, and potential fire behavior and effects least squares mean (standard error) values for the sampling areas. Values in the same row followed by a different letter are significantly different (Tukey's post-hoc ANOVA, p<0.05), p-values are for the ANOVA global F test. Sample size is 36 for all variables in each column except riparian reconstructed (n=35 for torching index, crowning index, and CBD), and upland reconstructed (n=35 for all variables). Fire behavior calculated under 97th weather conditions

	Lassen	Onion	W Tahoe	E Tahoe
BA (m^2/ha)	57.1(0.1)a	69.0(0.5)a	44.5(0.2)ab	31.3(0.3)b
Stand Dens(stems/ha)	370.5(1.1)a	308.8(1.2)a	310.7(1.1)a	299.4(1.2)a
QMD (cm)	44.4(2.5)a	50.7(4.8)a	46.1(2.9)a	38.1(3.9)a
Avg CBH (m)	7.5(0.01)a	8.5(0.02)a	6.6(0.01)a	6.3(0.02)a
% Comp (by BA)				
Fire Tolerant	13.3(0.3)ab	4.1(1.3)a	17.2(0.5)ab	42.5(0.9)b
Fire Intolerant	87.7(0.3)ab	95.9(1.3)a	82.8(0.5)ab	57.5(0.9)b
Snag Vol (m^3/ha)	6.5(0.3)a	10.1(0.6)a	9.5(0.3)a	4.7(0.5)a
Fuel Loads (Mg/ha)				
Duff	21.1(1.1)a	14.7(1.2)ab	14.5(1.1)ab	10.9(1.2)b
Litter	11.9(1.1)a	11.3(1.3)a	9.6(1.2)a	7.5(1.2)a
1 hr	1.1(1.2)a	0.9(1.4)ab	0.5(1.2)b	0.2(1.3)c
10 hr	2.3(1.1)a	1.9(1.3)ab	1.4(1.2)ab	1.1(1.2)b
100 hr	1.6(0.01)a	1.5(0.02)a	1.3(0.01)a	0.8(0.02)a
1000 hr	2.0(0.0)a	2.0(0.0)ab	1.2(0.0)ab	0.9(0.0)b
Total	64.2(1.1)a	51.5(1.2)ab	47.6(1.1)ab	33.3(1.2)b
Flame Length (m)				
Surface	0.8(1.1)a	0.4(1.3)b	0.5(1.1)b	0.6(1.2)ab
Crown	2.9(1.1)a	5.7(1.2)b	4.2(1.1)ab	3.8(1.2)ab
Prob of Torch	0.30(0.06)a	0.09(0.11)a	0.22(0.06)a	0.17(0.09)a
Torch Index (km/hr)	40.7(0.3)a	117.5(0.7)a	66.6(0.4)a	52.6(0.5)a
Crown Index (km/hr)	40.2(0.1)a	38.2(0.2)a	37.8(0.1)a	52.2(0.2)a
CBD (kg/m^3)	0.08(1.11)a	0.07(1.23)a	0.07(1.13)a	0.05(1.19)a
Mortality (% BA)	31.7(1.2)a	12.9(1.3)a	20.6(1.2)a	19.8(1.3)a

3.9. Riparian vs. Upland Canopy Cover, Seedling Density, and Shrub Cover

Notable differences emerged when riparian and upland canopy cover, seedling density, and shrub cover were analyzed by sampling area (Table 8). Across all four sampling areas combined, canopy cover was very similar between riparian and upland forests. The same pattern is evident for the Lassen, Onion Creek, and West Side Tahoe sampling areas. On the east side of the Tahoe Basin, however, canopy cover of upland forests is 22% lower on average than that of riparian forests (Table 8).

Table 8. Current riparian and upland mean canopy cover, seedling density, and shrub cover for the data set overall and by sampling area

	Riparian			Upland		
	Canopy Cover (%)	Seedlings (stems/ha)	Shrub Cover (%)	Canopy Cover (%)	Seedlings (stems/ha)	Shrub Cover (%)
Overall	54	3175	25	51	1812	10
Lassen	58	3333	11	59	2340	6
Onion Creek	53	5250	39	53	6250	6
West Side Tahoe	47	2209	30	46	433	12
East Side Tahoe	58	3167	41	36	60	18

Similarly, riparian seedling density was nearly twice as high that of upland forests for all study areas combined (Table 8). This pattern is largely driven by the East Side Tahoe sampling area, in which riparian seedling density is 53 times higher than upland seedling density. A similar trend of lower magnitude is noted in the West Side Tahoe and Lassen sampling areas, while the opposite trend is noted in Onion Creek, the most mesic of the sampling areas.

Shrub cover is generally higher in riparian than in upland forests for all sampling areas combined (Table 8). This trend holds when each sampling area is analyzed individually, although there is a strong gradient of differences. In Lassen, riparian shrub cover is 4% higher than upland shrub cover, while the difference in shrub cover is as high as 33% in Onion Creek. The east and west sides of the Tahoe Basin have intermediate differences between riparian and upland shrub cover of 23% and 18%, respectively.

4. Discussion

While this study compares current and reconstructed riparian and upland fire regimes and forest conditions, it does not imply that forests should be restored to reconstructed historical conditions, which may be neither feasible nor desirable in the context of altered anthropogenic influences and climatic conditions (Anderson and Moratto, 1996; Douglass and Bilbao, 1975; Millar and Woolfenden, 1999; Pierce et al., 2004; Rowley, 1985). A more effective restoration strategy may be to approximate the processes and conditions under which the target ecosystem evolved, which include frequent low-intensity fire in of the case of Sierran yellow pine and mixed conifer forests (generally occurring from 370 to 1700 m elevation in the northern Sierra, and from 760-2700 m in the southern part of the range) (Agee et al., 1978; Falk, 1990; Kilgore and Taylor, 1979; Parsons and DeBenedetti, 1979; SER, 1993; Vankat and Major, 1978). Rather than providing specific standards for restoring forests, the comparisons drawn in this study are intended to highlight the differential departure of current riparian and upland conditions from historic conditions, and offer a reference for the stand structure, species composition, and fuel loads produced by an active fire regime (Falk, 1990; White and Walker, 1997).

4.1. Fire Regimes

4.1.1. Riparian vs. Upland FRI

Riparian fire histories of the sites we sampled were very similar to their adjacent upland forests, but with a few important differences. At most sites, there was no significant difference between riparian and upland fire return intervals under the C1 and C10 filters, which fails to support our first hypothesis but is consistent with some studies comparing riparian and upland fire histories (Charron and Johnson, 2006; Olson and Agee, 2005). At these sites, the similarity of riparian and upland fire return intervals suggests that streams may not act as an effective buffer to fire activity and movement through the landscape. However, approximately one fourth of the sites did exhibit a significant difference between riparian and upland fire return intervals under both the C1 and C10 filters, similar to other studies of riparian fire history (Everett et al., 2003; Skinner, 2003), although riparian and upland FRI may not be directly comparable if the areas sampled are not of similar size. This suggests that riparian areas may reduce fire frequency and act as a buffer to fire movement in some cases, as proposed in our first hypothesis (Camp et al., 1997; Skinner and Chang, 1996; Taylor and Skinner, 2003).

At three sites, the riparian areas had significantly shorter fire return intervals than the upland areas, directly contradicting our first hypothesis and suggesting that riparian areas may have occasionally acted as a corridor for fire movement through the landscape (Dwire and Kauffman, 2003; Pettit and Naiman, 2007). Riparian zones typically exhibit higher soil moisture, and are thus often characterized by higher site quality than adjacent upland areas (Agee, 1998). This may result in more rapid rates of fuel production and, in some cases, more frequent fire return intervals in the riparian zone than in adjacent upland areas, where fuel may be limiting fire spread through the landscape. Additionally, these sites had extensive meadow systems associated with portions of the riparian area, which may have been centers of native American travel and use (Lindstrom et al., 2000; Olson and Agee, 2005). Numerous indigenous tribes in the Sierra Nevada used fire for a variety of purposes, and likely had an influence on fire regimes (Anderson and Moratto, 1996). The ferns, sedges and rushes common to extensive meadow systems were often used in Native American basketry, and would have been burned frequently to maintain their quality (Anderson, 2006). Areas which historically experienced heavy use by Native American populations are associated with shorter fire return intervals (Barrett and Arno, 1982), possibly due to the prevalence of anthropogenic ignitions, although the actual ignition source of most historic fires cannot be known and FRI calculated from different sample area sizes may not be directly comparable.

4.1.2. Riparian vs. Upland Seasonality

In both riparian and upland areas, a majority of the fire scars occurred during the dormant season (late summer to early fall in this region), which is consistent with other fire history studies in the Sierra Nevada (Moody et al., 2006; Stephens and Collins, 2004; Taylor and Beaty, 2005). Four of the six sites that exhibited a greater proportion of non-dormant season fires were in Jeffrey pine forest type, indicating that pine-dominated forests may have experienced more spring and early-summer fires than other forest types (Table 1). Although not significantly different, there is a trend of more early earlywood fire scars in upland areas, indicating that earlier season fires may be more prevalent in upland areas than in riparian areas, as proposed in our second hypothesis (Figure 3). This may result from riparian areas having cooler microclimates that retain snow longer into the summer drying period. These mesic conditions, which maintain higher fuel moisture into the late spring and early summer, possibly limit fire from spreading into riparian areas during the spring season.

4.1.3. Forest Characteristics

Riparian and upland fire return intervals are shorter in sites surrounded by upland forests with a high proportion of fire-tolerant pine species under both the C1 and C10 filters (>22.7% and >37.6%, respectively), suggesting that fires may move into riparian areas more easily in pine-dominated forests. Similarly, the NMS results for riparian C10 fires suggest a shorter fire return interval in riparian forests with a higher proportion of pine, supporting our third hypothesis. Although the CART results indicate that the shortest riparian C10 fire return intervals are in mixed-conifer forests, similarly short fire return intervals are found in Jeffrey-pine and white fir forest types. The association of shorter fire return intervals with pine-dominated sites has been well-demonstrated in numerous studies of upland forests (Gill and Taylor, 2009; McKelvey et al., 1996; Skinner and Chang, 1996; Stephens, 2001), and appears to hold true in some riparian forests as well.

4.1.4. Precipitation Regimes

Both riparian and upland fire return intervals of our east-side sites were some of the shortest we found amongst all our samples under both the C1 and C10 filters, supporting our fourth hypothesis. However, the fire return intervals of these sites were well within the range of our west-side sites. Similarity of east-side and west-side fire regimes by forest type have been documented in several studies (Gill and Taylor, 2009; North et al., 2009; Stephens, 2001; Taylor, 2004; Taylor and Beaty, 2005; Vaillant and Stephens 2009). All 7 of our east-side sites were in Jeffrey pine forest type, which tend to have shorter fire return intervals than other forest types. Furthermore, the eastern slope of the Sierra Nevada experiences a pronounced rain shadow effect, in which storms moving inland from the Pacific Ocean drop most of their precipitation west of the Sierra crest. The drier conditions of the east side may create more consistently favorable burning conditions (North et al., 2009), resulting in shorter fire return intervals.

4.1.5. Site Characteristics

Upland return interval appears to be shorter at higher elevation (>1709 m) under the C1 filter, appearing to contradict our fifth hypothesis. Similarly, the upland C10 NMS results show a trend of decreasing fire return interval with increasing elevation. However, the upland C10 regression tree indicates that fire return intervals are shorter at lower elevation (<1944 m), appearing to support our fifth hypothesis. This apparent contradiction is primarily driven by 6 sites in the CART analysis that could not be included in the NMS analysis because the fire scar record was insufficient for calculating some fire return interval metrics. Five of these sites are <1944 m elevation and have a mean upland C10 fire return interval <40 years. The tendency of fire return interval to increase with elevation has been demonstrated in some studies (Bekker and Taylor, 2001; Caprio and Swetnam, 1995; Gill and Taylor, 2009; Heyerdahl et al., 2001; Swetnam et al., 2000; Taylor, 2000), while others have suggested that forest type and stand isolation may be more important for determining fire return interval in some cases (North et al., 2009; Stephens, 2001). The trend of shorter C1 fire return intervals at higher elevation in our data may be explained by the large number of fire events that were recorded on only one tree per site at high elevations. This may be due to an increase in number of lightning strikes with elevation (van Wagtendonk and Cayan, 2008), which could result in many small fires each scarring only one tree. These fires may fail to spread due to sparse fuels, low fuel production rates (Agee et al., 1978; Stohlgren, 1988; Swetnam et al., 2000), higher fuel moisture, and lower fuel packing ratios (Albini, 1976; Martin et al., 1979; Rothermel, 1983; van Wagtendonk et al., 1998). Low elevations, conversely, provide fuel conditions favorable to greater fire rate of spread (Gill and Taylor, 2009), resulting in the pattern of longer C10 fire return intervals at lower elevations in our CART analysis.

4.1.6. Stream Characteristics

Under the C1 filter, riparian fire return intervals were shorter on more incised (width to depth ratio <6.3), smaller streams (depth <1.3 m), suggesting that wider, deeper streams may be more effective barriers to small-scale fire activity and spread in some cases. Similarly, riparian fire return intervals under the C10 filter are shorter on narrower streams with lower gradient, partially supporting (width, depth, width/depth ratio) and partially contradicting (gradient) our sixth hypothesis. Agee (1993) hypothesized that streams with wider riparian zones would experience longer fire return intervals than those with narrow riparian zones. While some studies found no significant difference between the fire-return intervals of small and large streams (Olson and Agee, 2005), others indicate that small headwater streams are influenced by fire to a greater degree than larger streams, which are influenced more by fluvial processes (Charron and Johnson, 2006). Similar studies comparing riparian fire return intervals have found that first order, high gradient streams in ravines have shorter fire return intervals than second and third order, low gradient streams in wide valleys (Everett et al., 2003; Skinner, 2003).

4.1.7. Fire-Climate Synchrony

Increased occurrence of fire in both riparian and upland forests across all sample sites was significantly correlated with drought cycles, as recorded in the

PDSI. This correlation was stronger in the upland areas than in the riparian areas, indicating that upland fire return intervals are more highly synchronized with summer drought conditions, seemingly contradicting our seventh hypothesis. The correlation between years of heightened regional fire activity in upland areas and PDSI dry years has been demonstrated in numerous fire history studies (Swetnam, 1993; Swetnam and Baisan, 2003; Taylor and Beaty, 2005), but has yet to be studied in riparian areas. Because riparian areas typically feature higher moisture and lower temperature conditions, they may be effective buffers to fire movement under all but the most severe drought conditions when their high fuel loads may permit higher severity fire than adjacent upland areas would experience (Dwire and Kauffman, 2003; Pettit and Naiman, 2007; Skinner and Chang, 1996).

4.1.8. Temporal Variability in FRI

While many fire history studies have recorded a sharp decline in fire frequency following Euro-American settlement (Beaty and Taylor, 2001; Beaty and Taylor, 2008; Caprio and Swetnam, 1995; Moody et al., 2006; Olson and Agee, 2005; Stephens, 2001; Stephens and Collins, 2004; Taylor, 2000; Taylor, 2004; Taylor and Skinner, 2003), most of our sample sites showed no significant difference in mean FRI before and after 1850, and four sites had a significantly shorter FRI after 1850. Although temporal analysis of FRI could not be conducted for some sites due to a lack of recorded fire events before 1850, many sites continue to record fire events well into the 20th century (Figures 2a and 2b). Patterns of pre-settlement fire frequencies continuing into the post-settlement (Scholl and Taylor, 2010) and even post-fire suppression (North et al., 2009) periods have been documented in some fire history studies. Our Lassen and Tahoe sampling areas experienced extensive railroad logging during the post-settlement period, which was accompanied by frequent slash fires (Lawson and Elliot, 2008; Lindstrom et al., 2000), which may explain the continuity, or decrease, in FRI at most of our sample sites. A database of fire perimeters in California shows that fires have continued to burn in these areas throughout the 20th and into the 21st century (FRAP, 2009), indicating that fires still occur in some of our sampling areas, even if fire regimes have been altered.

4.2. Stand Structure, Fuel Loads, and Fire Behavior

4.2.1. Current versus Reconstructed Forest Conditions

Overall, most of the reconstructed values for riparian and upland variables were within the range of variability described in other forest reconstructions, historic inventory data, and studies of forests with currently active fire regimes (i.e. recurrent fire at intervals similar to the range of variability found prior to EuroAmerican settlement) (Table 9). Variability was generally higher in riparian forests under both current and reconstructed conditions, except for reconstructed stand density, snag volume, probability of torching, torching and crowning indices, and mortality; current QMD, CBH, species composition, 10 and 100 hr fuel loads; and species composition, crown flame length, and CBD in both current and reconstructed stands, which had higher standard errors in upland forests (Table 4). The variability of riparian and upland forests is certainly subject to geographical variation, and caution should be taken when drawing generalizations about the differences between riparian and upland forests.

Both riparian and upland forests currently have significantly greater BA, stand density, snag volume, CBD, duff and total fuel load, and lower torching and crowning indices than their respective reconstructed conditions, supporting the ninth hypothesis. Additionally, current riparian stands have significantly higher potential surface and crown fire flame lengths, probability of torching, and mortality than reconstructed riparian stands, also supporting the ninth hypothesis. These trends in current versus historical stand structure are similar to those found in other reconstructions of historical Sierran coniferous forests, and comparisons with early 20th century forest inventory data (Bouldin, 1999; Lieberg, 1902; North et al., 2007; Scholl and Taylor, 2010; Sudworth, 1900; Taylor, 2004). While some studies have found that current BA is not significantly different from reconstructed BA in drier forest conditions (North et al., 2007; Taylor, 2004), other studies in more mesic conditions have found that current BA has approximately doubled since the time of the last fire, similar to the results of this study (Scholl and Taylor, 2010; Taylor, 2004).

Most studies have found that stand density has increased dramatically since the active fire period (i.e. the period of time when fires occurred at intervals within the range variation found prior to EuroAmerican settlement), by factors ranging from 3 to 33, which is a larger increase than is found in this study (North et al., 2007; Scholl and Taylor, 2010; Taylor, 2004). The trend of increasing stand density is corroborated by historical data suggesting that early 19th century Sierran coniferous forests had stem densities much lower than current conditions in this study (Bouldin, 1999; Lieberg, 1902; Sudworth, 1900).

Table 9. Reconstructed riparian and upland forest conditions compared with the range of variability under an active fire regime as described in existing literature

	Reconstructed		Range of	
	Riparian	Upland	Variability	Reference
BA (m^2/ha)	28.5	21.4	8.0-59.7	a, c, e, i, l, m, n
Stand Density (stems/ha)	207.7	201.1	16.2-280.0	a, c, e, i, l, m, n
QMD (cm)	40.0	38.4	33.0-67.5	e, i, l, m, n
Avg CBH (m)	6.5	6.3	4.9-6.1	a, c
% Composition				
Fire Tolerant	8.6	21.3	48.9-94.6	b, e, i, l, n
Fire Intolerant	91.4	78.7	5.4-51.1	b, e, i, l, n
Snag Density (snags/ha)	19.4	49.1	5.0-150.7	b, h, j, k, l
Fuel Loads (Mg/ha)				
Duff	3.3	3.0	NA	
Litter	8.8	6.9	0.4-23.9	k
1 hr	5.4	3.0	0.0-0.9	k
10 hr	6.3	5.5	0.0-7.0	k
100 hr	1.8	1.5	0.0-8.8	k
1000 hr	1.4	0.9	0.0-156.4	k
Total	27.9	22.3	0.4-183.7	f, k
Flame Length (m)				
Surface	0.4	0.5	1.0-2.0	c, f
Crown	0.9	0.4	7.1-9.2	g
Probability of Torching	0.03	0.08	NA	
Torching Index (km/hr)	176.3	98.6	22.0-67.0	c, g
Crowning Index (km/hr)	61.6	61.9	48.0-371.0	a, c, d, g
CBD (kg/m^3)	0.04	0.04	0.01-0.12	a, c, d, g
Mortality (% BA)	16.5	21.0	21.9	l

[a]Brown et al. (2008)
[b]Fule and Covington (1997)
[c]Fule et al. (2002)
[d]Fule et al. (2004)
[e]North et al. (2007)
[f]Ottmar et al. (1995)
[g]Roccaforte et al. (2008)
[h]Savage (1997)
[i]Scholl and Taylor (2010)
[j]Stephens (2000)
[k]Stephens (2004)
[l]Stephens et al. (2008)
[m]Stephens and Gill (2005)
[n]Taylor (2004)

Although it seems intuitive that the higher snag volumes in current riparian and upland stands are the result of the absence of frequent fires that would have historically consumed snags, it is possible that the reconstructions in this study failed to detect snags that were standing at the time of the last fire, but fell and decayed prior to data collection. Without other reconstructions or historical measurements of snag volume, it is impossible to determine whether the trend of increased snag volume is a real effect or an artifact of the reconstruction methods. However, snag densities in a current mixed-conifer forest with an active fire regime in northern Mexico are much lower than those in forests that have experienced fire suppression, suggesting that the absence of fire may indeed lead to increased snag density and volume (Barbour et al., 2002; Ganey, 1999; Savage, 1997; Stephens, 2004; Stephens and Finney, 2002).

The greater BA and stand density of current riparian and upland forests are reflected in the greater canopy fuels as well. The two- to three-fold increases in current riparian and upland CBD from reconstructed conditions are within the range of other comparisons between current and reconstructed stand conditions in the southwestern US and the Black Hills, which found increases ranging from 48% to 750% (Brown et al., 2008; Fulé et al., 2002; Fulé et al., 2004; Roccaforte et al., 2008). Similar trends occur for some classes of surface fuels, with duff loads increasing by an order of magnitude and total fuel loads approximately tripling from reconstructed to current conditions. Other comparisons of reconstructed and current total fuel loads found increases ranging from 2% to 43% in some river basins and watersheds, but decreases in others (Huff et al., 1995).

These changes in stand structure and fuels have made current riparian and upland forests more susceptible to high-intensity fire. Potential torching indices in current riparian and upland stands have decreased by 88% and 52%, respectively, which appears to be a greater change than the 39% to 66% declines in torching index found in other studies modeling current and reconstructed fire behavior (Fulé et al., 2002; Roccaforte et al., 2008). Similarly, potential crowning indices in current riparian and upland stands have decreased by 57% and 54%, respectively, which is within the range of the 23% to 86% declines in crowning index found in other studies (Fulé et al., 2002; Fulé et al., 2004; Roccaforte et al., 2008). The surface and crown fire flame lengths in current riparian forests have increased by 50% and 125%, respectively, which is less than the 134% to 515% increases in flame length predicted by other studies comparing potential fire behavior in current and reconstructed forests (Fulé et al., 2002; Roccaforte et al., 2008). The probability of torching in current riparian forests is 15 times that of reconstructed forests, and the predicted basal area mortality has increased from 16.5% to 30.6%. Observed fire-caused mortality in a forest with an active fire regime in northwestern Mexico was 21.8% which, when compared with the 40-95% mortality in fire-suppressed forests in southern California, reveals a similar trend (Franklin et al., 2006; Stephens et al., 2008). Differences in many stand structure and fuel load variables have resulted in current riparian and upland stands exhibiting greater potential for high-intensity fire than their reconstructed counterparts, supporting the ninth hypothesis.

However, differences in QMD, CBH, and 1 to 100 hr fuels appear to contradict the ninth hypothesis. This study was not designed to directly identify the mechanisms driving stand structure and fuel load differences, so we can only offer the following hypotheses as possible explanations. Upland forests currently have a significantly larger QMD than reconstructed upland stands, and riparian stands show a similar but non-significant trend, contrary to the significant decreases in QMD attributed to infilling of small trees observed in other reconstruction studies (Fule et al., 2002; North et al., 2007). In this study, the rapid growth of small trees in the absence of fire may result from highly productive site conditions and more than a century of growth between the reconstruction period and current measurements at many sites. Many of the trees aged (56%) were >40 cm DBH but <150 years old.

Higher upland CBH and lower 1 to 100 hr fuels were found in current forests than in the reconstructed stands. This may result from high stem densities in current stands and the delay between foliage and branch shedding as trees self-prune under low-light conditions (Fitzgerald, 2005). Many current stands have high canopy cover with the lower limbs of most trees dead and denuded of foliage. CBH, which measures distance from ground to green branches, was often high and duff fuel loads were very high. Fuel loads in the 1 to 100 hr classes were low possibly because trees had not yet begun shedding their dead lower limbs. The fuel accumulation equations used (van Wagtendonk and Moore, 2010) were developed in relatively low-density stands (average BA of 41.6 m^2/ha) that may be more representative of an active fire regime and conditions reconstructed in this study (FRAP, 2010), which have a greater input of woody fuels from fire-killed limbs.

4.2.2. Riparian versus Upland Forests, Current Conditions

Current riparian stands had significantly higher stem density, lower QMD, lower CBH, lower proportion of fire-tolerant species, higher probability of torching, and greater predicted mortality than current upland stands, supporting the tenth hypothesis. Current stem density was approximately 58% greater in riparian than upland stems, a trend similar to the observations of higher stem density closer to water bodies and stream channels in boreal (about four times greater), mixed-conifer, and pinyon pine forests (138% higher), although the opposite trend was found in a coastal Douglas-fir forest with hillslope stem density twice as high as that in the riparian areas (Harper and Macdonald, 2001; Russell and McBride, 2001; Segura and Snook, 1992; Wimberly and Spies, 2001). Highly productive riparian zones may be able to support greater infilling of small trees, resulting in a current QMD 17% lower than adjacent upland areas, a trend also found in pinyon pine and coastal Douglas-fir forests (Segura and Snook, 1992; Wimberly and Spies, 2001). While current average CBH in this study is nearly 3 m lower in the riparian forests than in the upland, visual assessment of vertical structure in drawn-to-scale illustrations of coastal Douglas-fir forests shows no trend in height to live crown with increasing distance from the stream channel (Poage, 1994).

The current proportion of species composition accounted for by fire-tolerant species was 16% greater in upland stands than riparian stands, which is consistent with findings of 13-52% greater prevalence of fire-tolerant species with increasing distance from the stream channel in coastal Douglas-fir forests (McGarigal and McComb, 1992; Nierenberg and Hibbs, 2000; Pabst and Spies, 1999; Wimberly and Spies, 2001). In boreal forests, however, the proportion of the more fire-tolerant balsam poplar (*Populus balsamifera ssp. balsamifera*) decreases relative to the less fire-tolerant quaking aspen as distance from the lakeshore increases (Harper and Macdonald, 2001). Similarly, prevalence of fire-intolerant conifers was more highly correlated with distance from the stream channel than prevalence of fire-tolerant conifers in a mixed-conifer forest, possibly indicating that upland forests may be more strongly associated with fire-intolerant than fire-tolerant species in some cases (Russell and McBride, 2001).

Denser riparian stands composed of primarily fire-intolerant species with more vertical continuity of canopy fuels may result in higher riparian fire severity. The doubling of the probability of torching and predicted mortality in current riparian stands compared to current upland stands found in this study is consistent with observations of greater occurrence of crown fire near stream channels in pinyon pine forests (Segura and Snook, 1992). In contrast, no difference in percent crown scorch between riparian and upland stands was found in mixed-evergreen, mixed-conifer, and ponderosa pine forest types of southwestern and northeastern Oregon (Halofsky and Hibbs, 2008). While other factors such as differences in topography between riparian areas and uplands may also influence fire behavior, differences in stand structure, composition, and potential fire behavior found in this study suggest that riparian forests currently may be more susceptible to high-intensity fire than upland forests, supporting the tenth hypothesis.

Analysis of the correlation between current upland and riparian variables suggests that currently there is greater similarity between adjacent riparian and upland stand structure than there was historically. Some stand structure variables such as BA and snag volume are significantly correlated under current conditions, but not under reconstructed conditions. Similarly, current average CBH is more highly correlated than reconstructed average CBH. This may be attributable to infilling of small trees facilitated by fire suppression, and accumulation of snags in the absence of an active fire regime (North et al., 2007; Stephens 2004).

While riparian and upland QMD were significantly correlated in reconstructed stands, the lack of significant correlation under current conditions may be the result of differential infilling of small trees due to differences in riparian and upland productivity (Camp et al., 1997; Olson and Agee, 2005; Segura and Snook, 1992; Skinner and Chang, 1996). The same productivity-driven differential infilling of fire-intolerant species in riparian areas may be responsible for the current non-significance of the correlation between riparian and upland species composition. In contrast, the significant correlation between riparian and upland species composition under reconstructed conditions may be associated with a higher proportion of fire-tolerant species across the landscape maintained by a historically active fire regime (North et al., 2007; Taylor, 2004). There is no consistent correlation between riparian and upland fuel classes under current or reconstructed conditions, suggesting that differences in productivity may drive fuel accumulation and decomposition on a site-specific basis, despite there being no significant difference between riparian and upland mean fuel loads. Increasing homogeneity in stand structure of adjacent riparian and upland forests may contribute to increased susceptibility to high-intensity fire across the landscape (Fulé et al., 2004), as evidenced by the higher correlation between riparian and upland surface flame lengths under current conditions.

In contrast, there appears to be less similarity between riparian and upland forests in other fire behavior variables under current compared to reconstructed conditions. Current riparian and upland crown fire flame length is not correlated, possibly reflecting greater susceptibility of riparian areas to high-intensity fire and torching (Segura and Snook, 1992). However, crown fire flame length was highly correlated for upland and riparian forests under reconstructed conditions, with both forests having low values. Similarly, while the probability of torching was more highly correlated between riparian and upland forests under reconstructed conditions with mostly low values, it is slightly less correlated under current conditions, perhaps due to the greater probability of torching in riparian areas. Finally, riparian and upland potential mortality was highly correlated, with predominately low values under reconstructed conditions, but is currently not significantly correlated due to increased predicted mortality in riparian forests under current conditions. Although homogeneity between riparian and upland forests may be increasing in some stand structure variables due to infilling of small trees, fire behavior appears to be diverging, with riparian forests becoming more susceptible to high-intensity fire.

Current differences between riparian and upland canopy cover, seedling density, and shrub cover can likely be attributed to the relative disparity in moisture conditions, which varies across sampling areas. The much higher canopy cover in riparian forests relative to upland forests on the east side of the Tahoe Basin are likely indicative of the difference in soil moisture between riparian and upland microclimates and soil moisture in this driest sampling area (DRI, 2009). Similarly, seedling density is more than an order of magnitude higher in east side Tahoe riparian forests than upland forests in the same sampling area, likely due to high moisture conditions in riparian areas relative to the much drier upland conditions on the east side (DRI, 2009). The opposite trend in seedling density for Onion Creek may represent the relative similarity of riparian and upland soil moisture in this wettest sampling area (DRI, 2009). Although the relationship between moisture conditions and the gradient of differences between riparian and upland shrub cover across sampling areas is more difficult to discern, the general trend of higher shrub cover in riparian areas can be attributed to the abundance of mesophytic shrubs such as *Alnus incana ssp. tenuifolia* and *Salix spp.* Although no attempt is made in this study to reconstruct historic canopy cover, seedling density, and shrub cover conditions, it is likely that differences between riparian and upland areas were generally of lower magnitude, given the similarity of riparian and upland fire regimes noted in this study.

4.2.3. Riparian versus Upland Forests, Reconstructed Conditions

There is no significant difference between reconstructed riparian and upland forests for the variables analyzed in this study (supporting the eleventh hypothesis), possibly due to the historical similarity of their fire regimes. Reconstructed upland fuel loads appear to be highly correlated with historic fire return interval, alluding to the fuel-driven occurrence of fire in these Sierran coniferous forest types (Jensen and McPherson, 2008). Interestingly, reconstructed riparian fuel loads are not highly correlated with FRI for any size classes, possibly suggesting a greater influence of weather conditions on fire occurrence. The significant correlation between reconstructed riparian CBD and FRI indicates that crown fuels accumulate uniformly with time since the last fire (Fulé et al., 2004), perhaps due to greater moisture availability in riparian zones. However, the fact that no other variables were significantly correlated with FRI suggests a great deal of heterogeneity in historic riparian and upland fire regimes at the landscape level.

CONCLUSION

Our study suggests that coniferous riparian forests in the Sierra Nevada historically experienced frequent fire, often at intervals similar to the adjacent upland forests. This relationship, however, does vary as a function of forest, site, stream and climate conditions. Managers should take into account local conditions when developing treatment prescriptions for riparian areas, considering how forest, site and stream characteristics would have likely influenced fire return intervals and subsequent fire effects. Riparian areas surrounded by forests with a high proportion of fire-tolerant pine species (about one third of the basal area or greater), especially those east of the Sierra crest, likely experienced more frequent fire than riparian areas in other forest types, and could be treated similarly to upland areas. Less intensive treatment, such as hand thinning and pile burning small trees, should be considered for riparian areas in other forest types. Riparian areas at higher elevation typically experienced longer fire return intervals under the C10 filter and therefore could be treated less intensively than the adjacent upland areas. Riparian areas at lower elevations could be treated similarly to upland areas. Riparian areas bordering small incised headwater streams historically experienced fire at frequencies similar to those of upland areas, and could thus be treated the same. Wider streams likely acted as an effective barrier to fire under some conditions, resulting in longer fire return intervals in adjacent riparian areas which could receive less intensive treatment than adjacent upland areas.

Results suggest that coniferous riparian forests in the northern Sierra Nevada historically had forest structure, composition, fuel loads, and fire behavior similar to adjacent uplands. However, both riparian and upland stands currently appear to be more fire prone than their historic conditions, with riparian areas significantly more so than adjacent upland areas. While active management of riparian forests is becoming more common (Holmes et al., 2010; Stone et al., 2010), riparian forests could be considered a high priority for restoration and fuel reduction treatments, with objectives similar to adjacent upland forests. If reintroduction of an active fire regime similar to historic conditions is desirable, treatments might focus on reducing basal area and stand density by removing small fire-intolerant tree species, and reducing surface fuel loads, especially the duff layer. Such treatments may reduce flame lengths, probability of torching, crowning index, and probability of mortality to their historic range of variability, which was likely similar for many adjacent riparian and upland forests. However, prescriptions should take local conditions such as species composition, precipitation regime, elevation, stream channel size and incision into account, which may have historically influenced the relationship between riparian and upland fire regimes. This will produce heterogeneity at the landscape scale, while restoring forests conditions that will facilitate resilience under changing climatic conditions.

Acknowledgements

We thank Bobette Jones and Coye Burnett of the Lassen National Forest; Dave Fournier, Victor Lyon and Sue Norman of the Lake Tahoe Basin Management Unit; Tamara Sasaki of California State Parks; Bill Champion of Nevada State Parks; Amy Cecchettini and Tina Carlsen of the California Tahoe Conservancy; Pam Lefrancois and Kathy Long of the North Tahoe Public Utility District; and Mark Rideout of the County of Placer Property Management Division for their support in facilitating the permitting process for conducting fieldwork on their respective land ownerships. Field data collection was miraculously completed with the help of Robert Brown, Matthew Eastlick, Darcy Kato, Lea Kromschroeder, Don Stevens, Brad Comito, and the Lassen National Forest monitoring crew. Lab work was accomplished only with the assistance of Esther Cole, Lyssa Dieffendorf, Rose Kormanyos, Lea Kromschroeder and Sky Steiner. Funding for this study was provided by the Lassen National Forest, the Southern Nevada Public Land Management Act, and the Pacific Southwest Research Station.

References

Agee, J.K., 1993. *Fire Ecology of Pacific Northwest Forests*. Island Press, Washington D.C.

Agee, J.K., 1998. *The landscape ecology of western forest fire regimes*. Northwest Sci. 72, 24-34.

Agee, J.K., Wakimoto, R.H., Biswell, H.H., 1978. *Fire and fuel dynamics of Sierra Nevada conifers*. For. Ecol. Manage. 1, 255-265.

Albini, F.A., 1976. *Estimating wildfire behavior and effects*. US For. Serv. Gen. Tech. Rep. INT-GTR-30.

Anderson, M.K., 2006. The use of fire by Native Americans in California, in: Sugihara, N.G., van Wagtendonk, J.W., Fites-Kaufman, J., Shaffer, K.E., Thode, A.E., (Eds.), Fire in California's Ecosystems. University of California Press, Berkeley, pp. 417-430.

Anderson, M.K., Moratto, J.J., 1996. *Native American land-use practices and ecological impacts, in: Sierra Nevada Ecosystem Project*: Final Report to Congress, Vol. II Assessments and Scientific Basis for Management Options, Wildland Resources Center Report No. 37. Centers for Water and Wildland Resources, University of California Press, Davis, pp. 187–206.

Arno, S. F., Sneck, K.M., 1977. *A method for determining fire history in coniferous forests of the mountain west.* US For. Serv. Gen. Tech. Rep. INT-GTR-42.

Barbour, M., Kelley, E., Maloney, P., Rizzo D., Royce, E., Fites-Kaufmann, J., 2002. *Present and past old-growth forests of the Lake Tahoe Basin, Sierra Nevada, U.S.* J. Veg. Sci. 13, 461-472.

Barrett, S.W., Arno, S.F., 1982. *Indian fires as an ecological influence in the northern Rockies*. J. For. 80, 647-651.

Beaty, R.M., Taylor, A.H., 2001. *Spatial and temporal variation of fire regimes in a mixed conifer forest landscape, Southern Cascades, California, USA*. J. Biogeogr. 28, 955-966.

Beaty, R.M., Taylor, A.H., 2008. *Fire history and the structure and dynamics of a mixed conifer forest landscape in the northern Sierra Nevada, Lake Tahoe Basin, California, USA*. For. Ecol. Manage. 255, 707-719.

Bekker, M.F., Taylor, A.H., 2001. *Gradient analysis of fire regimes in montane forests of the southern Cascade Range, Thousand Lakes Wilderness, California, USA.* Plant Ecol. 155, 15-28.

Berg, N.H., 1990. *Experimental forests and ranges: field research facilities of the Pacific Southwest Research Station.* US For. Serv. Gen. Tech. Rep. PSW-GTR-119.

Bouldin, J.R., 1999. *Twentieth-century changes in forests of the Sierra Nevada.* Ph.D. dissertation, University of California, Davis.

Bradshaw, L., McCormick, E., 2000. *FireFamily Plus user's guide, Version 2.0.* U.S. For. Serv. Gen. Tech. Rep. RMRS-GTR-067.

Breiman, L., Freidman, J.H., Olshen, R.A., Stone, C.J., 1984. *Classification and Regression Trees*. Chapman and Hall, New York.

Brown, J.K., 1974. *Handbook for inventorying downed woody material.* US For. Serv. Gen. Tech. Rep. INT-GTR-16.

Brown, P.M., Wienk, C.L., Symstad, A.J., 2008. *Fire and forest history at Mount Rushmore.* Ecol. Appl. 18, 1984-1999.

Camp, A., Oliver, C., Hessburg, P., Everett, R., 1997. *Predicting late-successional fire refugia pre-dating European settlement in the Wenatchee Mountains*. For. Ecol. Manage. 95, 63-77.

Caprio, A.C., Swetnam, T.W., 1995. *Historic fire regimes along an elevational gradient on the west slope of the Sierra Nevada, California, in: Brown, J.K., Mutch, R.W., Spoon, C.W., Wakimoto, R.H. (eds.), Proceedings on Fire in Wilderness and Park Management.*. US For. Serv. Gen. Tech. Rep. INT-GTR-320, pp. 8-14.

Charron, I., Johnson, E.A., 2006. *The importance of fires and floods on tree ages along mountainous gravel-bed streams*. Ecol. Appl. 16, 1757-1770.

Cook, E.R., 2000. *Nino 3 index reconstruction. International Tree-Ring Data Bank. IGBP PAGES/World Data Center-A for Paleoclimatology Data Contribution Series #2000-052.* NOAA/NGDC Paleoclimatology Program, Boulder, Colorado, USA. <http://www.ncdc.noaa.gov/paleo/recons.html>. Accessed 10 May 2009.

Cook, E.R., Meko, D.M., Stahle, D.W., Cleaveland, M.K., 1999. *Drought reconstructions for the continental United States.* J. Clim. 12, 1145-1162.

De'ath, G., 2002. *Multivariate regression trees: a new technique for modeling species environment relationships*. Ecology 83, 1105-1117.

Dieterich, J.H., 1980. *The composite fire interval – a tool for more accurate interpretation of fire history, in: Stokes, M.A., Dieterich, J.H. (eds.), Proceedings of the Fire History Workshop*. US For. Serv. Gen. Tech. Rep. GTR-RM-81, pp. 8-14.

Dieterich, J. H., Swetnam, T.W. 1984. *Dendrochronology of a fire scarred ponderosa pine.* For. Sci. 30, 238-247.

Dixon, G.E., 2002. *Essential FVS: a user's guide to the Forest Vegetation Simulator.* U.S. For. Serv. Internal Report.

Dolph, K.L., 1981. *Estimating past diameters of mixed-conifer species in the central Sierra Nevada*. USDA Pacific Southwest For. Range Exp. Stn. Res. Note PSW-353.

Douglass, W.A., Bilbao, J. 1975. *Amerikanuak: Basques in the New World*. University of Nevada Press, Reno.

Dunne, T., Leopold, L.B. 1978. *Water in Environmental Planning*. W.H. Freeman and Co., San Francisco.

DRI, 2009. *Western Regional Climate Center*. Accessed 31 March 2010. http://www.wrcc.dri.edu.

Dwire, K.A., Kauffman, J.B., 2003. *Fire and riparian ecosystems in landscapes of the western USA*. For. Ecol. Manage. 168, 71-74.

Everett, R., Schellhaas, R., Ohlson, P., Spurbeck, D., Keenum, D., 2003. *Continuity in fire disturbance between riparian and adjacent sideslope Douglas-fir forests*. For. Ecol. Manage. 175, 31-47.

Falk, D.A., 1990. *Discovering the future, creating the past: some reflections on restoration. Restor.* Manage. Notes 8,71-72.

FEMAT (Forest Ecosystem Management Assessment Team), 1993. *Forest Ecosystem Management: An Ecological, Economic, and Social Assessment: Report of the Forest Ecosystem Management Team, July 1993: USDA, Forest Service; USDC, National Marine Fisheries Service; USDI, Bureau of Land Management, Fish and Wildlife Service, and National Park Service;* Environmental Protection Agency. U.S. Government Printing Office, Portland. FRAP (Fire and Resource Assessment Program), 2009. Fire Perimeters. Accessed 6 March 2010. http://frap.cdf.ca.gov/data/frapgisdata/select.asp.

Fitzgerald, S.A., 2005. *Fire ecology of ponderosa pine and the rebuilding of fire-resilient ponderosa pine ecosystems*. U.S. For. Serv. Gen. Tech. Rep. PSW-GTR-198.

Franklin, J., Spears-Lebrun, L.A., Deutschman, D.H., Mardsen, K., 2006. *Impact of a high-intensity fire on mixed evergreen and mixed conifer forests in the Peninsular Ranges of southern California*. For. Ecol. Manage. 235, 18-29.

Fulé, P.Z., Covington, W.W., 1997. *Fire regimes and forest structure in the Sierra Madre Occidental, Durango, Mexico*. Acta Bot. Mex. 41, 43-79.

Fulé, P.Z., Covington, W.W., Moore, M.M., 1997. *Determining reference conditions for ecosystem management of southwestern ponderosa pine forests*. Ecol. Appl. 7, 895-908.

Fulé, P.Z., Covington, W.W., Smith, H.B., Springer, J.D., Heinlein, T.A., Huisinga, K.D., Moore, M.M. 2002. *Comparing ecological restoration alternatives: Grand Canyon, Arizona*. For. Ecol. Manage. 170, 19-41.

Fulé, P.Z., Crouse, J.E., Cocke, A.E., Moore, M.M., Covington, W.W. 2004. *Changes in canopy fuels and potential fire behavior 1880-2040: Grand Canyon, Arizona*. Ecol. Model. 175, 231-248.

Ganey, J.L. 1999. *Snag density and composition of snag populations on two National Forests in northern Arizona*. For. Ecol. Manage. 117, 169-178.

Gill, L., Taylor, A.H., 2009. *Top-down and bottom-up controls on fire regimes along an elevational gradient on the east slope of the Sierra Nevada, California, USA*. Fire Ecol. 5:57-75.

Grissino-Mayer, H.D., 2001. FHX2: *Software for the analysis of fire history from tree rings*. Tree-Ring Res. 57, 113-122.

Grissino-Mayer, H.D., Swetnam, T.W. 2000. *Century-scale climate forcing of fire regimes in the American Southwest.* Holocene 10, 213-220.

Harmon, M.E., Cromack, K., Jr., Smith, B.G., 1987. *Coarse woody debris in mixed-conifer forests, Sequoia National Park, California*. Can. J. For. Res. 17, 1265-1272.

Harper, K.A., Macdonald, S.E., 2001. *Structure and composition of riparian boreal forest: new methods for analyzing edge influence*. Ecol. 82, 649-659.

Heyerdahl, E.K., Brubaker, L.B., Agee, J.K., 2001. *Spatial controls of historical fire regimes: a multiscale example from the interior west, USA*. Ecology 82, 660-678.

Heyerdahl, E.K., McKenzie, D., Daniels, L.D., Hessl, A.E., Littell, J.S., Mantua, N.J., 2008. *Climate drivers of regionally synchronous fires in the inland Northwest (1651-1900).* Int. J. Wildland Fire 17, 40-49.

Holmes, S.B., Kreutzweiser, D.P., Hamilton, P.S., 2010. *Operational and economic feasibility of logging within forested riparian zones*. Forestry Chronicle 86, 601-607.

Huff, M.H., Ottmar, R.D., Alvarado, E., Vihnanek, R.E., Lehmkuhl, J.F., Hessburg, P.F., Everett, R.L. 1995. *Historical and current forest landscapes in eastern Oregon and Washington.* U.S. For. Serv. Gen. Tech. Rep. PNW-GTR-355.

Innes, J.C., North, M.P., Williamson, N. 2006. *Effects of thinning and prescribed fire restoration treatments on woody debris and snag dynamics in a Sierran old-growth, mixed-conifer forest.* Can. J. For. Res. 36, 3183-3193.

Jensen, S.E., McPherson, G.R. 2008. *Living with Fire.* University of California Press, Berkeley.

Keane, R.E., 2008. *Biophysical controls on surface fuel litterfall and decomposition in the northern Rocky Mountains, USA*. Can. J. For. Res. 38, 1431-1445.

Kilgore, B.M., Taylor, D. 1979. *Fire history of a sequoia – mixed conifer forest.* Ecology 60, 129-142.

Kimmey, J.W., 1955. *Rate of deterioration of fire-killed timber in California*. U.S. Dep. Agric. Circ. 962.

Kitzberger, T., Brown, P.M., Heyerdahl, E.K., Swetnam, T.W., Veblen, T.T. 2007. *Contingent Pacific-Atlantic ocean influence on multi-century wildfire synchrony over western North America*. Proc. Natl. Acad. Sci. U.S.A. 104, 543-548.

Lawson, S.J., Elliot, D.R. 2008. *Logging in Plumas County*. Arcadia Publishing, San Francisco.

Lieberg, J.B. 1902. *Forest conditions in the northern Sierra Nevada, California.* U.S. Geological Survey Prof. Pap. 8, Series H, Forestry 5, Government Printing Office, Washington, D.C.

Lindstrom, S., Rucks, P., Wigand, P., 2000. *A contextual overview of human land use and environmental conditions, in: Murphey, D.D., Knopp, C.M. (eds.), Lake Tahoe Watershed Assessment: Volume II.* U.S. For. Serv. Gen. Tech. Rep. PSW-GTR-175.

Martin, R.E., Anderson, H.E., Boyer, W.D. 1979. *Effects of fire on fuels: a state-of-knowledge review.* US For. Serv. Gen. Tech. Rep. GTR-WO-13.

Mast, J.N, Fulé, P.Z., Moore, M.M., Covington, W.W., Waltz, A.E.M. 1999. *Restoration of presettlement age structure of an Arizona ponderosa pine forest.* Ecol. Appl. 9, 228-239.

McCaffery, S., Moghaddas, J.J., Stephens, S.L. 2008. *Survey results from fire and fire surrogate fuel treatments in a Sierran mixed conifer forest, California, USA.* Int. J. Wildland Fire 17, 224-233.

McCune, B., Grace, J.B. 2002. *Analysis of Ecological Communities.* MjM Software Design, Gleneden Beach.

McGarigal, K., McComb, W.C. 1992. *Streamside versus upslope breeding bird communities in the central Oregon coast range.* J. Wildl. Manage. 56, 10-23.

McKelvey K.S., Busse, K.K. 1996. *Twentieth century fire patterns on Forest Service lands, in: Sierra Nevada Ecosystem Project, Final Report to Congress, Assessments and Scientific Basis for Management Options,* Vol. II, Centers for Water and Wildland Resources, University of California Press, Davis, pp. 1119–1138.

McKelvey, K.S., Skinner, C.N., Chang, C., Erman, D.C., Husari, S.J., Parsons, D.J., van Wagtendonk, J.W., Weatherspoon, P.C. 1996. *An overview of fire in the Sierra Nevada, in: Sierra Nevada Ecosystem Project: Final report to Congress, Vol. II, Assessments and Scientific Basis for Management Options.* Centers for Water and Wildland Resources, University of California Press, Davis, pp.1033-1040.

McKenzie, J.D., Jr., Goldman, R. 1999. *The student edition of Minitab for Windows user's manual: release 12.* Addison-Wesley, New York.

Millar, C.I., Woolfenden, W.B. 1999. *The role of climate change in interpreting historical variability.* Ecol. Appl. 9, 1207-1216.

Moody, T.J., Fites-Kaufman, J., Stephens, S.L. 2006. *Fire history and climate influences from forests in the northern Sierra Nevada, USA.* Fire Ecol. 2, 115-141.

Moore, B.M., Lee, B.G., Davey, S.M. 1991. *A new method for predicting vegetation distributions using decision tree analysis in a geographic information system*. Environ. Manage. 15, 59-71.

Moore, M.M., Huffman, D.W., Fulé, P.Z., Covington, W.W., Crouse, J.E. 2004. *Comparison of historical and contemporary forest structure and composition of permanent plots in southwestern ponderosa pine forests*. For. Sci. 50, 162-176.

Morrison, M.L., and Raphael, M.G. 1993. *Modeling the dynamics of snags*. Ecol. Appl. 3, 322-330.

Murphy, K., Rich, T., Sexton, T. 2007. *An assessment of fuel treatment effects on fire behavior, suppression effectiveness, and structure ignition on the Angora Fire*. US For. Serv. Tech. Pap. R5-TP-025.

Nierenberg, T.R., Hibbs, D.E. 2000. *A characterization of unmanaged riparian areas in the central coast range of western Oregon*. For. Ecol. Manage. 129, 195-206.

Norman, S.P., Taylor, A.H. 2003. *Tropical and north Pacific teleconnections influence fire regimes in pine-dominated forests of Northeastern California, USA*. J. Biogeogr. 30, 1081-1092.

North, M., Innes, J., Zald, H. 2007. *Comparison of thinning and prescribed fire restoration treatments to Sierran mixed-conifer historic conditions*. Can. J. For. Res. 37, 331-342.

North, M.P., Van de Water, K.M., Stephens, S.L., Collins, B.M. 2009. *Climate, rain shadow, and human-use influences on eastern Sierra Nevada fire regimes.* Fire Ecol. 5, 20-34.

Oliver, C.D., and Larson, B.C. 1996. *Forest stand dynamics*. Updated ed. John Wiley and Sons, New York.

Olson, D.L., Agee, J.K. 2005. *Historical fires in Douglas-fir dominated riparian forests of the southern Cascades, Oregon*. Fire Ecol. 1, 50-74.

Pabst, R.J., Spies, T.A. 1999. *Structure and composition of unmanaged riparian forests in the coastal mountains of Oregon*, U.S.A. Can. J. For. Res. 29, 1557-1573.

Parsons, D.J., DeBenedetti, S.H. 1979. *Impact of fire suppression on a mixed-conifer forest.* For. Ecol. Manage. 2, 21-33.

Pettit, N.E., Naiman, R.J. 2007. *Fire in the riparian zone: characteristics and ecological consequences*. Ecosystems 10, 673-687.

Pierce, J.L., Meyer, G.A., Jull, A.J.T. 2004. *Fire-induced erosion and millennial-scale climate change in northern ponderosa pine forests*. Nature 432, 87-90.

Poage, N.J. 1994. *Comparison of stand development of a deciduous-dominated riparian forest and a coniferous-dominated riparian forest in the Oregon coast range.* Master's thesis, Oregon State University, Department of Forest Science, Corvallis.

Pyne, S.J. 1982. *Fire in America: A Cultural History of Wildland and Rural Fire.* Princeton University Press, Princeton.

Rambo, T., North, M. 2008. *Spatial and temporal variability of canopy microclimate in a Sierra Nevada riparian forest.* Northwest Sci. 82, 259-268.

Rambo, T., North, M. 2009. *Canopy microclimate response to pattern and density of thinning in a Sierra Nevada forest.* For. Ecol. Manage. 257, 435-442.

Raphael, M.G., and Morrison, M.L. 1987. *Decay and dynamics of snags in the Sierra Nevada, California.* For. Sci. 33, 774-783.

Reinhardt, E., Crookston, N.L. 2003. *The Fire and Fuels Extension to the Forest Vegetation Simulator.* U.S. For. Serv. Gen. Tech. Rep. RMRS-GTR-116.

Roccaforte, J.P., Fule, P.Z., Covington, W.W. 2008. *Landscape-scale changes in canopy fuels and potential fire behavior following ponderosa pine restoration treatments.* Int. J. Wildland Fire 17, 293-303.

Rothermel, R.C. 1983. *How to predict the spread and intensity of wildfires.* US For. Serv. Gen. Tech. Rep. GTR-INT-143.

Rowley, W.D. 1985. U.S. *Forest Service grazing and rangelands.* Texas AandM University Press, College Station.

Russell, W.H., McBride, J.R. 2001. *The relative importance of fire and watercourse proximity in determining stand composition in mixed conifer riparian forests.* For. Ecol. Manage. 150, 259-265.

Safford, H.D., Schmidt, D.A., Carlson, C.H. 2009. *Effects of fuel treatments on fire severity in an area of wildland-urban interface, Angora Fire, Lake Tahoe Basin, California.* For. Ecol. Manage. 258, 773-787.

Savage, M. 1997. *The role of anthropogenic influences in a mixed-conifer forest mortality episode.* J. Veg. Sci. 8, 95-104.

Scholl, A.E., Taylor, A.H. 2010. *Fire regimes, forest change, and self-organization in an old-growth mixed-conifer forest,* Yosemite National Park, USA. Ecol. Appl. In Press.

Segura, G., Snook, L.C. 1992. *Stand dynamics and regeneration patterns of a pinyon pine forest in east central Mexico.* For. Ecol. Manage. 47, 175-194.

Shapiro, S.S., Wilk, M.B. 1965. An analysis of variance test for normality (complete samples). Biometrika 52, 591-611.

Skinner, C.N. 2003. *A tree-ring based fire history of riparian reserves in the Klamath Mountains, in: Faber, P.M. (ed.), California Riparian Systems: Processes and Floodplains Management, Ecology, and Restoration, Riparian habitat and floodplains conference proceedings.* March 12-15, 2001, Sacramento, CA. Riparian Habitat Joint Venture, Sacramento, pp. 116-119.

Skinner, C.N., Burk, J.H., Barbour, M., Franco-Vizcaino, E., Stephens, S.L. 2008. *Longer-term influences of climate on fire regimes in montane forests of northwestern Mexico*. J. Biogeogr. 35, 1436-1451.

Skinner, C.N., Chang, C. 1996. *Fire regimes, past and present, in: Sierra Nevada Ecosystem Project, Final Report to Congress, Vol. II:* Assessments and Scientific Basis for Management Options. Center for Water and Wildland Resources, University of California Press, Davis, pp. 1041-1069.

SER (Society for Ecological Restoration). 1993. *Mission statement*. Restor. Ecol. 1, 206-207.

Stephens, S.L. 2000. *Mixed conifer and upper montane forest structure and uses in 1899 from the central and northern Sierra Nevada. California Madrono* 47, 43-52.

Stephens, S.L. 2001. *Fire history differences in adjacent Jeffrey Pine and upper montane forests of the eastern Sierra Nevada*. Int. J. Wildland Fire 10, 161-167.

Stephens, S.L. 2004. *Fuel loads, snag abundance, and snag recruitment in an unmanaged Jeffrey pine-mixed conifer forest in Northwestern Mexico*. For. Ecol. Manage. 199, 103-113.

Stephens, S.L., Collins, B.M. 2004. *Fire regimes of mixed conifer forests in the north-central Sierra Nevada at multiple spatial scales*. Northwest Sci. 78, 12-23.

Stephens, S.L., Finney, M.A. 2002. *Prescribed fire mortality of Sierra Nevada mixed conifer tree species: effects of crown damage and forest floor combustion*. For. Ecol. Manage. 162, 261-271.

Stephens, S.L., Gill, S.J. 2005. *Forest structure and mortality in an old-growth Jeffrey pine-mixed conifer forest in northwestern Mexico*. For. Ecol. Manage. 205, 15-28.

Stephens, S.L., Fry, D.L., Franco-Vizcaino, E. 2008. *Wildfire and spatial patterns in forests in northwestern Mexico: the United States wishes it had similar fire problems*. Ecol. Soc. 13 (2), 10 [online] URL: http://www.ecologyandsociety.org/vol13/iss2/art10/

Stephens, S.L., Martin, R.E., Clinton, N.E. 2007. *Prehistoric fire area and emissions from California's forests, woodlands, shrublands and grasslands*. For. Ecol. Manage. 251, 205-216.

Stephens, S.L., Moghaddas, J.J. 2005. *Experimental fuel treatment impacts on forest structure, potential fire behavior, and predicted tree mortality in a mixed conifer forest*. For. Ecol. Manage. 215, 21-36.

Stohlgren, T. 1988. *Litter dynamics in two Sierran mixed conifer forests I. Litter fall and decomposition rates*. Can. J. For. Res. 18, 1127-1135.

Stokes, M.A., Smiley, T.L. 1977. *An Introduction to Tree-ring Dating*. University of Chicago Press, Chicago.

Stone, K.R., Pilliod, D.S., Dwire, K.A., Rhoades, C.C., Wollrab, S.P., *Young, M.K. 2010. Fuel reduction management practices in riparian areas of the western U.S.A.* Env. Manage. 46, 91-100.

Strahler, A.N. 1952. *Hypsometric (area-altitude) analysis of erosional topology.* Geol. Soc. Am. Bull. 63, 1117-1142.

Sudworth, G.B. 1900. *Stanislaus and Lake Tahoe Forest Reserves, California, and adjacent territory.* U.S. Geological Survey Prof. Pap. 8, Series H, Forestry 5. Government Printing Office, Washington, D.C.

Swetnam, T.W. 1993. *Fire history and climate change in sequoia groves.* Science 262, 885-889.

Swetnam, T.W., Baisan, C.H. 1996. *Historical fire regime patterns in the southwestern United States since AD 1700, in: Allen, C. (ed.), Fire Effects in Southwestern Forests,* Proceedings of the Second La Mesa Fire Symposium, Los Alamos, New Mexico, March 29-31, 1994. US For. Serv. Gen. Tech. Rep. RM-GTR-286, pp. 11-32.

Swetnam, T.W., Baisan, C.H. 2003. *Tree ring reconstructions of fire and climate history in the Sierra Nevada and southwestern United States,* in: Veblen, T.T., Baker, W., Montenegro, G., Swetnam, T.W. (eds.), Fire and Climatic Change in Temperate Ecosystems of the Western Americas. Springer-Verlag, New York.

Swetnam, T.W., Baisan, C.H., Morino, M., Caprio, A.C. 2000. *Fire history along elevational transects in the Sierra Nevada, California, in: Final report to Sierra Nevada Global Change Research Program*, United States Geological Survey, Biological Resources Division, Sequoia, Kings Canyon, and Yosemite National Parks. University of Arizona Press, Tucson.

Taylor, A.H. 2000. *Fire regimes and forest changes in mid and upper montane forests of the southern Cascades, Lassen Volcanic National Park, USA.* J. Biogeogr. 27:87-104.

Taylor, A.H. 2004. *Identifying forest reference conditions on early cut-over lands, Lake Tahoe Basin,* USA. Ecological Applications 14, 1903-1920.

Taylor, A.H., Beaty, R.M. 2005. *Climatic influences on fire regimes in northern Sierra Nevada mountains*, Lake Tahoe Basin, Nevada, USA. J. Biogeogr. 32, 425-438.

Taylor, A.H., Skinner, C.N. 2003. *Spatial patterns and controls on historical fire regimes and forest structure in the Klamath Mountains*. Ecol. Appl. 13, 704-719.

USDA. 2004. *Sierra Nevada Forest Plan Amendment Final Supplemental Environmental Impact Statement.* US For. Serv., Pacific Southwest Region.

U.S. Forest Service. 1993. *National Interagency Fire Management Integrated Database reference manual.* U.S. For. Serv., Washington, D.C.

U.S. Forest Service. 1996. *Kansas City Fire Access Software user's guide*. U.S. For. Serv., Washington, D.C.

Vaillant, N.M., Stephens, S.L. 2009. *Fire history of a lower elevation Jeffrey pine mixed-conifer forest in the eastern Sierra Nevada, California, USA*. Fire Ecol. 5, 4-19.

Vankat, J.L., Major, J. 1978. *Vegetation changes in Sequoia National Park, California.* J. Biogeogr. 5, 377-402.

van Wagtendonk, J.W., Benedict, J.M., Sydoriak, W.M. 1998. *Fuel bed characteristics of Sierra Nevada conifers.* West. J. Appl. For. 13:73-84.

van Wagtendonk, J.W., Cayan, D.R. 2008. *Temporal and spatial distribution of lightning strikes in California in relation to large-scale weather patterns*. Fire Ecol. 4:34-56.

van Wagtendonk, J.W., Moore, P.E. 2010. *Fuel deposition rates of montane and subalpine conifers in the central Sierra Nevada, California, USA.* For. Ecol. Manage. 259, 2122-2132.

Vayssieres, M.P., Plant, R.E., Allen-Diaz, B.H. 2000. *Classification trees: an alternative nonparametric approach for predicting species distributions.* J. Veg. Sci. 11:679-694.

Wensel, L.C., Olson, C.M. 1995. *Tree taper model volume equations*. Hilgardia 62, 1-73.

White, P.S., Walker, J.L. 1997. *Approximating nature's variation: selecting and using reference information in restoration ecology*. Restor. Ecol. 5,338-349.

Wimberly, M.C., Spies, T.A. 2001. *Influences of environment and disturbance on forest patterns in coastal Oregon watersheds*. Ecol. 82, 1443-1459.

Wykoff, W.R., Crookston, N.L., Stage, A.R. 1982. *User's guide to the Stand Prognosis Model.* U.S. For. Serv. Gen. Tech. Rep. INT-GTR-133.

In: Riparian Zones
Editor: A. Kerem et al.
ISBN 978-1-61942-677-1

Chapter 2

ECOLOGICAL AND SOCIOECONOMIC STAKES FOR CONSERVING AND PROTECTING RIPARIAN FORESTS IN THE DAHOMEY GAP (WEST AFRICA)

Kouami Kokou*[*1], Armand Natta[2], Kossi Adjonou[1] and Adzo Dzifa Kokutse[1]

[1]Laboratoire de Botanique et Ecologie Végétale, Faculté des Sciences, Université de Lomé, Togo

[2]Faculté d'Agronomie, Université de Parakou, Parakou, Bénin

ABSTRACT

Riparian forests are vegetation along waterways both in savannah and forest area. In the Dahomey Gap which is the interruption in the West African forest block at the level of Togo and Benin, riparian forests are the main frequently visible forests. In this part of Africa devoid of natural forests, riparian forests play a fundamental role in the conservation of biodiversity, are involved in the ecological functioning of streams, improve water quality, contribute to the bank stabilization of rivers and reduce the impact of sewage and chemical spills from cities. They are also of major socio-economic importance for local populations. These forests along the

* corresponding author: 1Laboratoire de Botanique et Ecologie Végétale, Faculté des Sciences, Université de Lomé, Togo ; Email: kokoukouami@hotmail.com

rivers are still relatively protected by the local populations but intensive agriculture, illegal tree cutting, and charcoal production due to the high demographic growth rate in the region seriously impacted them nowadays. Scientifically, there is an important lack of information specifically relating to the riparian forests of the zone, in which new plant species are regularly assessed. Currently, the potential importance of these riparian forests in the local forestry projects in progress in Togo and Benin, and their patrimonial value for re-afforestation programs, justify the need to set up a monitoring system to get a better knowledge from them. Therefore, this study aim to "document" 1) the physical, social, economic and cultural conditions in which these forests continue to exist, 2) their roles in conservation and protection of habitats and biodiversity, 3) current threats which if left unchecked will lead to their disappearance and, 4) ecological indicators, policies and legal tools to advocate for sustainable management. These information are useful to exploit the potential importance and the phytogenetic value of these riparian forests for monitoring and the local forestry projects.

Keywords: Dahomey Gap, Togo, Benin, Riparian forests, socio economic and ecological roles.

1. Introduction

In Togo and Benin, located inside the interruption of the West-African forest belt called Dahomey Gap (DG) (White 1983, IUCN 1992), riparian forests (RFs) are the most perceptible forests. With the disappearance of the relics of forest ecosystems in the southern part of both countries, RFs become the only witness of forest plant communities. Abiotic characteristics of the RFs and their heterogeneity, associated to the difficulty to exploit them, contribute to support a higher biological diversity (Bowman and McDonough 1991, Bornette and Amoros 1996, Bornette et al. 1996, Tabacchi and Planty-Tabacchi 1996, Bornette et al. 1998, Bendix and Hupp 2000, Da Costa Filho et al. 2006, Malm Renofalt et al. 2007). In addition, RFs play a role in the ecological functioning of watercourses (Décamps and Naiman 1989), improve water quality, contribute to the riverside stability and a better dissolution of oxygen in the water, reduce the consequence of waste water and chemical flows, decline water temperature, etc. (FAO 2003).

Throughout many areas in the world, temperate or subhumid tropical habitats, authors present RFs as the refuges for flora i.e. ecosystems which preserve the main part of the hotspots of biodiversity (Meave et al. 1991, Meave and Kellman

1994). They support a high level of diversity compared to non-RFs in the same zones (Nilsson et al. 1989, Baker 1990, Tabacchi et al. 1990, Medley 1992, Meave and Kellman 1994, Ferreira 1997, Pither and Kellman 2002, Goebel et al. 2003, Natta 2003). Tabacchi et al. (1990) identified 900 plant species along Adour River in France.

Unfortunatelly these forests along the rivers which are relatively saved by the local populations some few years ago, are nowadays impacted by agriculture and charcoal production due to the high demographic growth rate in the DG. Despite the pressure, the potential importance of the RFs in the local forestry projects currently in progress in Benin and Togo, and their patrimonial value, justify the need to get better knowledge of them. To reach this objective, this paper synthesises the knowledge available in both countries, highlighting the role of these riverside forests in the conservation of biodiversity, their socio economic and cultural importance as well as the threats they are facing in order to provide databases for monitoring system and reporting on their sustainable management.

2. Environmental and Socio Economic Status of Riparian Forest in the DG

The RFs studied are located throughout Bénin and Togo territories in West Africa, between latitudes 6°N and 12°25'N and longitudes 0°E and 3°55'E (Figure 1). In general, the profile of these countries is an undulating plateau except for Atacora chain though Togo from southwest to north-east and in the north of Benin and few scattered hills in the centre and the north of both countries. The altitude varies from sea level to 400-900 m, where the Atacora chain is the outstanding feature. Four major groups of soils can be distinguished: (1) ferrallitic soils covered by semi-deciduous forest, (2) ferruginous soils covered by dry forest, woodland and savannah, (3) vertisols in the depression of Lama covered by savannahs and patches of dry semi-deciduous forest, and (4) hydromorphic soils covered by swamp and RFs.

The climate is primarily determined by the annual cycle of the "Inner Tropical Convergence Zone" (ITCZ). Three climate zones can broadly be distinguished: (1) From the coast up to 7°–7°30'N, the climate is subequatorial with two rainy seasons alternating with a long dry season (December-February), and a short dry season (July-August), which rarely exceeds two months; (2) Between the latitudes 7° and 9°, the climate becomes subhumid or subsudanian with a tendency to a pattern of one rainy season and one dry season; (3) The northern section is characterized by a truly Sudanian climate with a unimodal rainfall regime. The mean annual rainfall varies from 800 to 1300 mm. The lowest value (800 mm) is recorded in the southwest of Togo at the level of the capital town Lomé. The highest precipitation is confined in the southern part of Atakora Mountains in Togo characterized by a transitional subequatorial climate with a long rainy season occurring from March to October and disrupted by a slight decrease in rainfall in August or September. The rainfall amounts range from 1300 to 1600 mm per year. Here, semi-deciduous moist forests are the major vegetation type. The mean annual temperature ranges from 26 to 28°C and can exceptionally reach 35-40°C in northern localities such Dapaong in Togo or Malanville in Bénin. The annual temperature amplitude is low (5-10°C) in the south while it is high (11-13°C) in the north (from the latitude 8°N northwards).

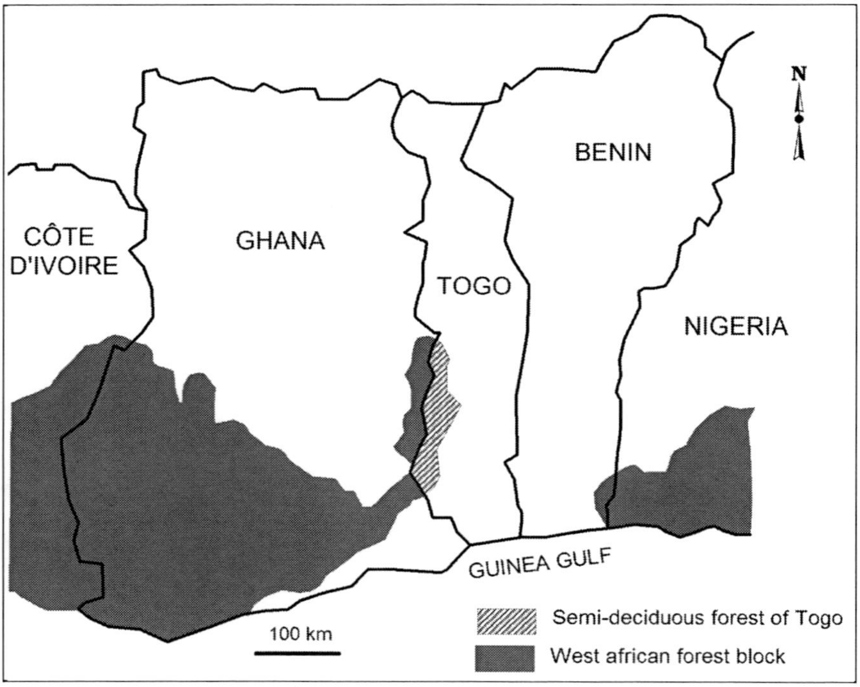

Figure 1. Location of Benin and Togo in the Dahomey Gap.

The very developped hydrographic network exploits the structural or tectonical directions of the region (Figure 2). It is constituated of a series of dense rivers, more or less important, divided into three hydrographic basins in Togo 1) Volta Basin; 2) Mono Basin with Anié and Amou Sub basin; 3) Zio Basin. In

Benin there are four major hydrographic basins: 1) the high and upper parts of the Ouémé Basin in South and Central Benin; 2) the Niger Basin and its tributaries i.e. the Mekrou, Alibori and Sota in North Benin; 3) the upper part of the Pendjari Basin in North West Benin; 4) and the Basin of the Mono river in south West Benin, that is prolonged in South East Togo.

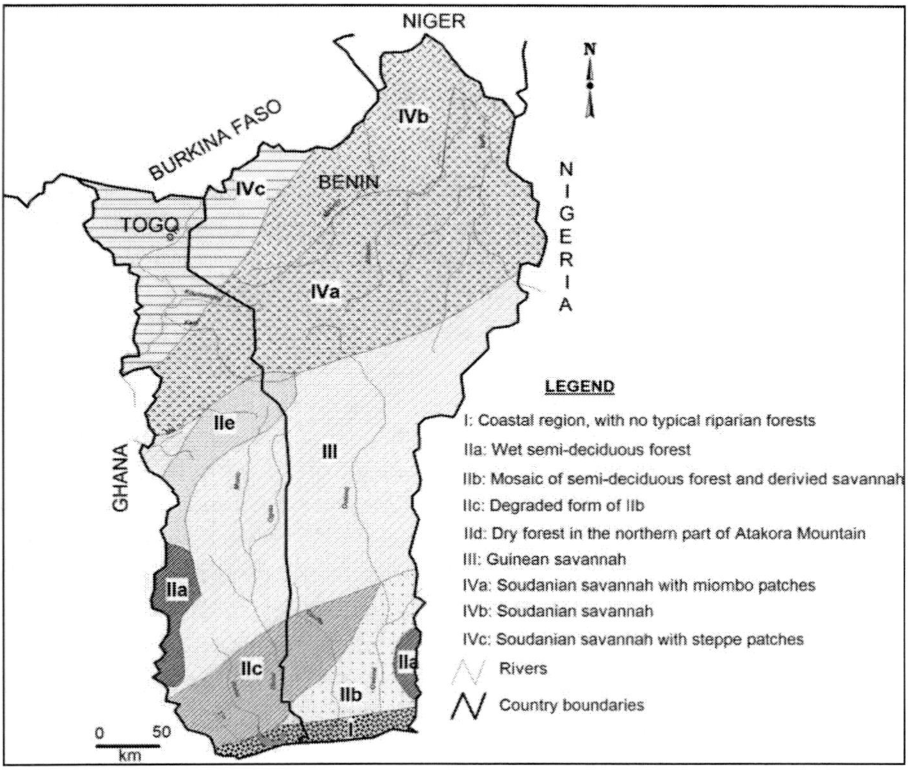

Figure 2. Main vegetation types and watercourses in Benin and Togo (Adapted from Adjanohoun et al. 1986, 1989).

Benin and Togo belongs to the DG, which is the dry wedge that separates the West African rain forest belt into the Upper Guinea and Lower Guinea/Congolian Forest Blocks (White 1983). This part of the West Africa consists of savannah, grassland, farmland, and fallow intermingled with small islands of closed forest (semi deciduous and swamp forests) (Figure 2). From 7°30' N to 12°25' N, the natural vegetation consists essentially of a patchwork of woodlands and savannahs with belts of RFs along rivers.

The patches of RFs survive in a densely populated area with 250 to 300 hab/km^2 in the southern region and 15 to 20 hab/km^2 in the northern region. Towns growing, extensive farming in rural areas, cattle ranching and forest products harvesting compromise the maintining of the RFs.

3. Major Ecological Challenges

3.1. Conservation and Protection of Habitats and Biodiversity

3.1.1. Conserving Threatened Habitats in RFs

Phytogeographical linkages were established between RFs and sempervirent forests located at the west and east of the DG. In these forests, the climate is relatively humid in the major part (1600 to 2000 mm/an) and the monthly temperature is about constant throughout the year. Approximately 8000 plant species were recorded inside these forests, among which many are endemic (Jones 1955, Guillaumet and Adjanohoun 1971, Hall and Swaine 1981). The RFs studied in the south of Atakora Mountain (Adjossou 2004) were also described in the forests of Ghana (Hall and Swaine 1981) and Côte d'Ivoire (Guillaumet and Adjanohoun 1971). In southeast Benin, domain of dense semi-deciduous forests (Adomou 2011) and the continuation of the lowland rain forest zone of Nigeria (Jones 1955), the RFs described by Natta et al. (2003) were found in the forest types of Nigeria and Cameroun-Gabon forest area (Keay et al 1964, Letouzey 1985). RFs in the centre of Benin and Togo show that they are the prolongation of dry semi-deciduous forestand fire sub-type of northern Guinean zone (Keay 1960, Hall and Swaine 1981, Adomou 2011). Likewise, north Benin prolongs the sudanian zone of Nigeria (Keay 1949) and the plains of northern Togo (Ern 1979).

Some of the main conserved habitats in the RFs are:

3.1.1.1. In Forest Zone

The coastal region of southern Benin and Togo (I, cf Legend, Figure 2) is the domain of marshlands, lakes, lagoons, coastal thickets or schrubs and a number of small relics of sacred groves. The present-day pattern of vegetation is very much obscured by the omnipresent impacts of dense human populations. This flat region lacks typical riparian forests. However, in the Pobè phytogeographic district (South Est Benin) that is the domain of dense semi-deciduous forests (IIa, IIb, IIc, Figure 2), Sokpon (1995) identified a riverine forest dominated by *Cleistopholis patens, Ficus mucuso, Elaeis guineensis* and *Cercestis mirabilis* in the protected forest of Pobè. The equivalent of this phytogeographic area in Togo is a farmland area (IIc, Figure II) where along Haho, Zio and Mono rivers. In this area, three main plants communities were identified according to geomorphology and soil texture (Kokou and Caballé 2000):

- *Cleistopholis* patens RFs closed to the meridional area of Atakora Mountains (IId Figure 2). The characteristic species are *Cleistopholis patens*, *Pycnanthus angolensis*, *Myrianthus serratus* and *Rhigiocarya racemiflora*;
- *Pterocarpus santalinoides* RFs in the valley of Zio. In these forest galleries, two plant species are exclusive i.e. are found only there in Togo; they are *Dichapetalum oblongum* and *Placodiscus attenuatus;*
- *Diospyros mespiliformis* RFs in the Haho valley, covering a large area where *Elaeis guineensis* is very frequent.

In the humid forest zone in the south of Atakora Moutains (IIa, Figure 2), where semi-deciduous forests are almonst disappearing under the influence of coffee and cocoa farming, Adjossou (2004) described many forest types conserved along the watercources. The main plant communities are:

- *Parinari glabra* RFs on the plateaus of Danyi and Adélé. These RFs were established on the drained river banks at altitude generally up to 500 m. Trees in these RFs are 25 to35 mhigh;
- *Pentadesma butyraccea* RFs on the Akposso plateau. They were settled in clay soils, relatively deep and drained. Generally these RFs are located at the savannahs and forests boundaries but could follow the watercourses until the source;
- *Pterocarpus mildbraedii-Celtis mildbraedii* RFs observed on the Adélé plateau at 600 m of altitude in a large depression on a very deep clay-sand soil. Trees were about 40 m high;
- *Spondiathus preussii* RFs located on Akébou and Adelé plateaus along river banks on reddish sandstone soils containing patches of iron oxide. *Berlinia* grandiflora is often associated to *Spondiathus preussii* in these RFs*;*
- *Cathormion altissimum* RFs found as well in moutain as in the plain througout this area;
- *Pentaclethra macrophylla* are RFs type of half and high slope (altitude less to 500 m);

- *Pterocarpus santalinoides* RFs growing in the flat bottom of depression or along large stream with stagnant water during the low water period.

3.1.1.2. In Guinean Savannahs

In the Guinean savannahs (III, Figure 2) and in the dry forests in the northern part of Atakora Mountain (IId, Figure 2), RFs follow most of the watercourses. Therefore they develop a strip of dense forests throughout the landscape of dry forests and savannahs. Along the Ouémé River in southern Benin, Natta *et al.* (2002) identified four plant communities:

- *Cynometra megalophylla* and *Parinari congensis* RFs with dense canopies, 20 to 30 high. Common species in this plant community are *Napoleonaea vogelii, Cassipourea congoensis, Sallacia pallescens, Synsepalum brevipes, Cleistopholus patens, Pseudospondias microcarpa, Pierrodendron kerstingii, Spathodea campanulata, Trilepiseum madagascariense, Parquetina nigrescens*, etc. ;
- *Mimosa pigra* and *Ficus asperifolia* RFs that occur as dense shrubby vegetation on sandy bars.
- *Capparis thonningii* and *Crataeva adansonii* RFs along the Ouémé River in the soudano-guinean region;
- *Lepisanthes senegalensis* and *Drypetes floribunda* along the Ouémé River in the soudano-guinean region.

Along Ouémé and its tributaries at the level of Penessoulou protected forest, two main plant communities described by Natta *et al.* (2002) are:

- *Isolona thonneri* and *Callichilia barteri* along streams in the central part;
- *Motandra guineensis* and *Pararistolochia goldieniana* at the east and west parts.

In the Plateau and Couffo phytogeographical districts (Adomou 2011) which prolong respectively the derived savannahs (Jones 1945) and the Southern Guinea zones of Nigeria (Keay 1959), the dominant tree species in the RFs are *Pterocarpus santalinoides, Cola laurifolia, Parinari congensis, Antidesma venosum, Napoleonaea vogelii, Syzygium guineense, Dialium guineense* and *Cynometra megalophylla*.

The Zou district is a transition zone between the Guinean and Sudanian climates. Riparian forests phytosiognomy is similar to the dense semi-deciduous forests, and the most frequent species are *Berlinia grandifolia, Elaeis guineensis, Hexalobus crispiflorus, Poutera alnifolia, Cola gigantea, C. millenii, Lecaniodiscus cupanioides, Napoleonaea vogelii, Pterocarpus santalinoides* and *Uvaria chamae*.

The Bassila region has particular features of dense semi-deciduous forest type (Hall and Sawaine 1981). Here the most frequent plants species in RFs are *Pterocarpus santalinoides, Khaya grandifoliola, K. senegalensis, Pentadesma butyracea, Isolona thonneri, Callichili barteri* and *Motandra guineensis*.

In the Guinean zone of Togo, the major characteristics of RFs are synthesized as follow:

- *Milicia excelsa* RFs common in Fazao-Malfakassa Park or in the gazzetted forest of Alédjo. Apart *Milicia excelsa*, two main species more represented are *Ceiba pentandra* and *Manilkara multinervis*. Close to the river bed, *Pseudospondias microcarpa* and *Dacryodes klaineana* are frequent. The undergrowth is dominated by *Hypselodelphys poggeana*. These RFs could be considered as the relics of Moraceae and Sapotaceae described by Akpagana (1989);
- *Diospyros mespiliformis* and *Manilkara multinervis* RFs occupy the tributaries of the rivers. Two bigest trees which characterize the canopy are *Anogeissus leiocarpus* and *Borassus aethiopum*. Sometime, each of these two species constitutes a monospecific population. The low stratum is dominated by *Lecaniodiscus cupanioides* and *Malacantha alnifolia*. Other frequent species in these galleries are *Cola gigantea, Millettia thonningii, Napolaeona vogelii, Polysphaeria arbuscula, Diospyros mobuttensis,* etc. The big trees are often covering by lianas such as *Saba comoriensis*. In the open area, *Chromolaena odorata* invade completely the undergrowth;
- *Breonadia salicina* and *Cassipourea congensis* RFs are found in dry forest area. *Diospyros mespiliformis, Manilkara multinervis, Anogeissus leiocarpus, Syzygium guineense, Lonchocarpus sericeus*, etc. are the most big trees and *Pouchetia africana, Linociera nilotica, Polysphaera arbuscula* the main shrub. *Saba comoriensis, Strychnos* sp., *Uvaria chamae, Opilia amentacea, Taccazzea apiculata* constitute the lianas. Palm tree (*Borassus aethiopum, Raphia* sp., *Phoenix reclinata*) are abundant.

In the upstream of Nangbeto Dam on Mono River, major RFs are:

- Plant community of the minor bed of the River with main taxa as *Cynometra megalophylla*, *Cola laurifolia*, *Xylopia aethiopica*, *Ficus capraefolia*, *Mimosa pigra*, *Phyllanthus reticulatus*. The last three taxa constitute dense ticket on the sandy bars;
- Plant community of the tributaries with *Diospyros mespiliformis* and *Manilkara multinervis* dominated the gallery, relayed in places by *Millettia thonningii*. In the undergrowth, *Polysphaeria arbuscula* and *Oxyanthus speciosus* are the most frequent shrubs. In the bed on the block of mother rocks were developing different species of ferns;
- The contact savannah-gallery is indicated by the species as *Andropogon tectorum* for the Graminae and *Anogeissus leiocarpus* for the woody plant.

3.1.1.3. In Soudanian Savannahs

RFs or gallery forests in the savannahs region of north Benin and Togo (IVa, IVb, IVc, Figure 2) are observed along tributaries of the Rivers, streams and brooks. They are classified according to the stream order, longitude, latitude, relief and topography and the seasonnality of water flow:

- RFs along large river beds: generally they were constituted of three fringes parralel to the river bed :

*Fringe 1*_is developing on the ten first meters of the bank. More frequent sepecies are *Cola laurifolia, Combretum micranthum, Diospyros mespiliformis, Parinari congensis, Pterocarpus santalinoides*, etc. (Photo 1);

Fringe 2, between 10 and 20, some 30 m according to the width of teh RFs, with species like *Anogeissus leiocarpus, Diospyros mespiliformis, Oncoba spinosa, Dichrostachys cinerea, Kigelia africana*, etc. ;

Fringe 3 is in contact with the savannah or the dry forest that extends the gallery forest, with species like *Uvaria chamae, Tamarindus indica, Anogeissus leiocarpus*. This last fringe is under bush fire influence every year.

- RFs along rivers tributaries: the waterflow dries up completely in these streams during the low water period. The main species are *Anogeissus leiocarpus, Combretum micranthum, Diospyros mespiliformis, Oncoba spinosa, Tamarindus indica, Uvaria chamae,* etc.
- RFs along stream with early low water : These streams form meanders within the savannahs, making the floristic composition very heterogenous

with main species like *Khaya senegalensis*, *Vitex doniana*, *Terminalia glaucescens*, *Combretum nigricans,* etc.

Photo 1. RF along Koumongou River in the Oti-Keran National Park (Togo).

In the sudanian zone of North Benin, many plant communities were discriminated in the gallery forests (Sokpon et al. 2001, Natta et al. 2002). They are:

- *Chrysobalanus icaco* subsp. *atacoriensis* and *Pentadesma butyracea* along streams at hill feet in the Atacora mountain chain;
- *Alchornea cordifolia* and *Ficus trichopoda* along streams on regularly inundated plateaus all over the country;
- *Berlinia grandiflora* and *Khaya senegalensis* along streams on drained plateaus (*i.e.* seldom inundated);
- *Raphia sudanica* and *Oxytenanthera abyssinica* along streams on drained plateaus;
- *Uapaca heudelotii* and *Irvingia smithii* along the Sota River in the north-east of Benin;
- *Breonadia salicina* and *Garcinia ovalifolia* RFs at the foot of the hills with permanent water flow. The species are from dense semi deciduous forest types with 15 to 25 m high;

- *Cassia sieberiana* and *Terminalia glaucescens* RFs on drained and seasonal streams on plateaus. The canopy is less luxuriant and the gallery forest is similar to woodland savanna surrounded by shrub or tree savanna;
- *Cola laurifolia* and *Parinari congensis*, characteristic of stream in the north east of the Pendjari Parc. Trees are 10 to 16 m high and inundated all along the rainy season;
- *Garcinia livingstonei* and *Combretum acutum* along Magou stream in the north west of the Pendjari Park.

3.1.2. Conserving Biodiversity Hotspots in the RFS

Floristic survey showed that RFs contain about 1,200 vascular plant species, about at least 40% of the countries estimated flora. These plant species were divided into 79% dicotyledons, 16% monocotyledons, 450 genus and 110 families. The four most important families by their number of species were Rubiaceae (70 species), Fabaceae (41), Euphorbiaceae (36) and Moraceae (31). Eleven families had between 14 and 10 species, 20 families between 9 and 5 species and 58 families were represented per less than 5 species (Figure 3). The preponderance in Rubiaceae is comparable with the other wet tropical forests (Akoègninou 1984, Akpagana 1989, Kolongo et al 2004, Kouamé et al. 2004, Langenberger et al. 2006, Tchouto et al. 2006, Ssegawa and Nkuutu 2006). For example, in the wet forests of Ivory Coast, Rubiaceae species dominate with 260 species on a total of 2,126 inventoried species (Kouamé et al. 2004). In the small islands of dense wet forests in the Republic of Benin, the family of Rubiaceae contains the greatest number of species (Akoègninou 1984).

The most common species in RFs in both countries are *Pterocarpus santalinoides, Mimosa pigra, Ficus asperifolia, Dialium guineense, Elaeis guineensis, Cola laurifolia, Berlinia grandifolia, Parinari congensis, Kigelia africana, Crataeva adansonii, Morelia senegalensis, Syzygium guineensis var. guineensis, Alchornea cordifolia, Khaya senegalensis, Uapaca heudelotii* (Brunel et al 1984, Natta et al. 2002, Kokou et al. 2008). In general all over Benin and Togo, apart from national parcs and their hunting zones, RFs are treathened by human activities.

With this plant diversity, RFs can be considered as hotspots of biodiversity for several reasons. They are among the most vulnerable forest vegetations at all latitudes, yet of high ecological importance because they are vital ecosystems for numerous wild and bird life species; they protect many water resources; harbour numerous plant species either rare, threatened, or with superior adaptability to a specific habitat. Moreover they are known for their high species richness on relatively small patches and narrow corridors.

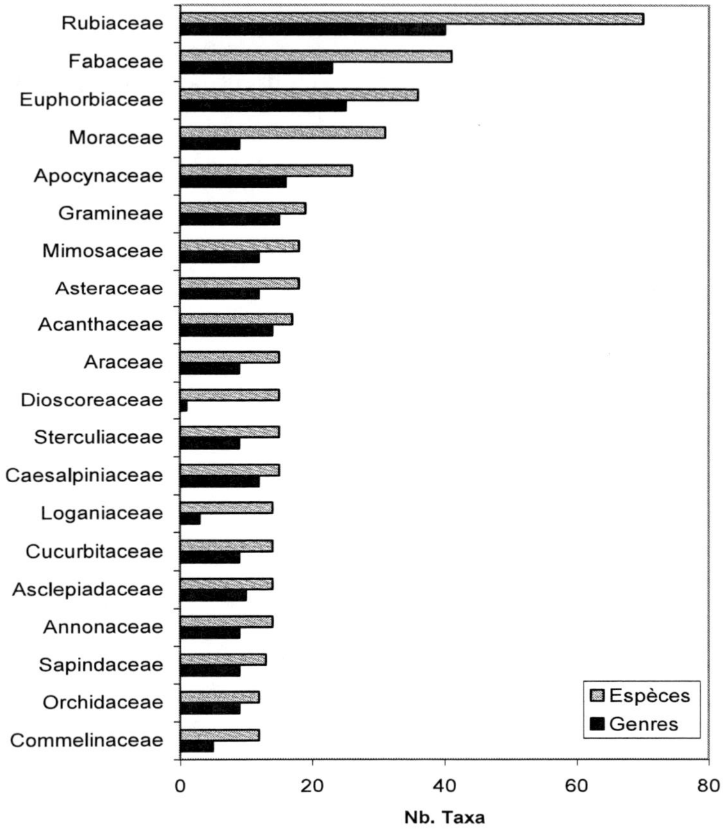

Figure 3. Most common plant families in the RF in the DG.

In addition to being characterized by a diversity of plant species, and unique assemblage of plants, compared to uplands and wetlands, 44 species in RFs were endemic of the African western forest block. In Togo, 40 species recorded in RFs were not known in the flora before (Table 1). Endemic genus of the African western block of forests identified by Kouamé et al. (2004) such as *Amphimas, Aubrevillea, Discoglypremna, Distemonanthus, Hymenostegia,* and *Turraeanthus* were observed in RFs. In Benin a newly described rare and endemic ornamental Acanthaceae, *Thunbergia atacorensis*, is found along certain streams at Atakora hills feet in a fire-prone environment (Natta 2003). Such occurrence is not negligible and cannot be underestimated within the Dahomey Gap. Moreover *Cercopithecus erythrogaster erythrogaster*, the endemic primate found so far in Benin (Oates 1996, Grubb et al. 1999, Sinsin et al. 2002), shares a part of only its time in RFs and forest patches in the Ouémé valley region.

Table 1. Some plant species quoted for the first time by studing RFs in the DG.

N°	Family	Scientific name	Life form
1	Acanthaceae	*Thunbergia atacorensis*	Shrub
2	Araceae	*Culcasia lanceofolia*	Veine
3	Asteraceae	*Mikaniopsis metlandii*	Veine
4	Annonaceae	*Isolona cooperi*	Shrub
5	Boraginaceae	*Cordia alliodora*	Tree
6	Boraginaceae	*Cordia platythyrsa*	Tree
7	Ebenaceae	*Diospyros elliotii*	Schrub
8	Euphorbiaceae	*Drypetes aylmeri*	Schrub
9	Euphorbiaceae	*Uapaca esculentus*	Tree
10	Euphorbiaceae	*Protomagabaria stoptefana*	Schrub
11	Fabaceae	*Dalbergia adami*	Veine
12	Flacourtiaceae	*Caloncoba sp.*	Schrub
13	Flacourtiaceae	*Dovyalis zenkeri*	Schrub
14	Flacourtiaceae	*Flacourtia vogelii*	Schrub
15	Flacourtiaceae	*Caloncoba wilwitschii*	Tree
16	Flacourtiaceae	*Caloncoba echinata*	Tree
17	Hippocrateaceae	*Salacia erecta*	Veine
18	Moraceae	*Ficus anomani*	Tree
19	Melastomataceae	*Dissotis multiflora*	Herb
20	Orchidaceae	*Nervilea subintegra*	Herb
21	Orchidaceae	*Polystachya niodesta*	Herb
22	Moraceae	*Ficus tesmanii*	Tree
23	Pandaceae	*Panda oleosa*	Tree
24	Passifloraceae	*Smeathmannia laevigata*	Schrub
25	Rubiaceae	*Tricalysia chevaleri*	Schrub
26	Rubiaceae	*Tricalysia paniculata*	Schrub
27	Rubiaceae	*Uncaria taboltri*	Veine
28	Sapindaceae	*Pancovia turbinata*	Schrub
29	Sapotaceae	*Chrysophyllum africanum*	Tree

The forest species identified in the RFs in the DG represent 20% of the 3,660 species of the total flora of the Ivory Coast (Heywood and Davis 1994), 43% of the 1,700 species of the flora Sierra Leone (Heywood and Davis 1994), and 23% of the 2,800 species of the vascular plants listed within the African western forest block (Jongkind 2004). The comparison between the floristic list of the RFs in the DG and species previously recorded in the threatened natural forest fragments in Benin and Togo (Akoègninou 1984, Akpagana 1989), shows that the basic elements (88.5%) are preserved in the RFs. The species of the original flora are thus preserved in the RFs. This result approaches the observations made in the RFs of the Mount Pine Ridge in Balize in Central America (Meave et al. 1991, Meave and Kellman 1994, Pither and Kellman 2002).

The distribution of the species by biological life form showed that the phanerophyts accounted for 82% of the biological spectrum and constituted the basic component of the flora of RFs, i.e. 34% microphanerophyts, 25% mesophanerophyts, 18% nanophanerophyts and 5% megaphanerophyts. The hemicryptophyts (4.98%) and geophyts (4.98%) had relatively higher frequeny compared to the chamephyts, epiphyts, therophyts and hydrophyts, respectively 2.90%, 2.49%, 1.66% and 0.97%.

The distribution of the species by phytogeographical origin showed that taxa from the guineo-congolese area (wet dense forest) were most frequent with 64% of the listed species, of which 56% were common to the western and eastern guineo-congolese forests, 7% were specific to the western forest (West of the DG) and 1% to the eastern forest (East of the DG). The species of the soudano-zambezian area accounted only for 5%. Taxa common to the guineo-congolese and soudano-zambezian zones occupied also a relatively high proportion (26%). Introduced and cultivated species in RFs accounted for 5%. Among 220 very rare species recorded in the RFs, 117 (25.43%) were typical of wet tropical forests, 66 (34.37%) occupied the wet tropical forests and other habitats, 25 (67.56%) were not associated to forest habitats and 12 (35.29%) were introduced species. Twenty eigth percent of the forest species were rare in the RFs. The common epiphyts were particularly rare; on 18 taxa listed 11 were rare.

3.1.3. Ecological Importance, Soils and Water Protection

All RFs relics are to be protected for their ecosystems services and valuable goods. A part the importance of RFs as critical, seasonally or temporary habitat for a variety of species, the fundamental and critical ecological functions that RFs perform better than uplands and aquatic ecosystems, fall into four major categories: (a) hydrology and sediments dynamics, (b) biogeochemistry and nutrient cycling, (c) species and habitats diversity, and d) food web maintenance.

More than other vegetation types, they are well-known for their roles in controlling/regulating water runoff and quality, nutrient flows, bank erosion and stabilisation, sedimentation, flood, air and soil pollution, water shading and cooling, sediment filtering, aquifer recharge, carbon storage, foliage inputs, biomass storage, solar radiation (Forman and Godron 1986, Gregory et al. 1991, Harper et al. 1994, Fetherston et al. 1995, Ettema et al. 1999, Giese et al. 2003).

Maintaining biodiversity is one of the most important functions of RFs, which are, directly or indirectly, home to an abundance of animal life (e.g. fishes, birds, invertebrates and micro-organisms, amphibians, reptiles, mammal species). The reproduction, growth and nutrition of many animals depend on the presence of more or less dense and safe stands of RFs lining waterways and inundation plains (e.g. nesting and living in trees; food consisting of fish, (semi) aquatic animals, algae, invertebrates, and micro organisms).

The characteristic plant species, plant communities and associated aquatic or semi-aquatic animal species in RFs are intrinsically linked to the role of water as both agent of natural disturbance and as critical requirement of biota survival. In savannah regions where large tracks of dense forests does not exist, RFs play the role of residual forest and provide, when they are undisturbed, the best conditions for wildlife that normally lives in the forest zone.

4. Socio Economic and Cultural Roles of RFs

The timber potential of the RFs is high; about 400 woody species were recorded inside RFs. The reference to the list of timber exploited in Benin and Togo enable to know that about 40 species were regularly harvested in the RFs (Table 2). Leguminosae occupy the half of the exploited timber. The same situation were observed in the forests in Congo Basin (Cameroon, Congo, Gabon, Central Africa) where the Leguminosae constitute a group particularly important by the number of species and the volume exploited (Détienne 1991). Apart of the timber, 100% of populations in rural areas use wood as combustible. They also supply cities in firewood.

Some of this riverside forests are sacred, in which case legends mark their origin, practices of the local communities inside exist, the nature of the idol that generates cultural and food taboos also. All these customs wrigths are the reason that these forest still exist today as habitats of gods or ancestors, thereby having different social and cultural functions than other forests.

Table 2. Economical value timbers of RFs

N°	Family	Scientific name
1	Anacardiaceae	*Lannea kerstingii*
2	Annonaceae	*Cleistopholis patens*
3	Apocynaceae	*Alstonia boonei*
4	Bombacaceae	*Ceiba pentandra*
5	Burseraceae	*Canarium schweinfurthii*
6	Chrysobalanaceae	*Parinari glabra*
7	Combretaceae	*Terminalia superba*
8	Ebenaceae	*Diospyros mespiliformis*
9	Euphorbiaceae	*Uapaca guineens*is
10	Euphorbiaceae	*Uapaca heudelotii*
11	Fabaceae	*Afzelia africana*
12	Fabaceae	*Berlinia grandiflora*
13	Fabaceae	*Daniella thurifera*
14	Fabaceae	*Detarium senegalense*
15	Fabaceae	*Dialium guineense*
16	Fabaceae	*Distemonanthus benthamianus*
17	Fabaceae	*Erythrophleum suaveolens*
18	Fabaceae	*Amphimas pterocarpoides*
19	Fabaceae	*Erythrina mildbraedii*
20	Fabaceae	*Pterocarpus mildbraedei*
21	Guttiferae	*Symphonia globulifera*
22	Meliaceae	*Khaya grandifoliola*
23	Meliaceae	*Khaya senegalensis*
24	Meliaceae	*Cedrela odorata*
25	Meliaceae	*Trichilia heudelotii*
26	Meliaceae	*Turraeanthus africana*
27	Mimosaceae	*Albizia adianthifolia*
28	Mimosaceae	*Albizia ferruginea*
29	Mimosaceae	*Albizia glaberrima*
30	Mimosaceae	*Albizia zygia*
31	Mimosaceae	*Aubrevillea kerstingii*
32	Mimosaceae	*Pentaclethra macrophylla*
33	Mimosaceae	*Piptadeniastrum africanum*
34	Moraceae	*Antiaris africana*
35	Moraceae	*Milicia excelsa*
36	Myristicaceae	*Pycnanthus angolensis*
20	Simaroubaceae	*Hannoa klaineana*
37	Sterculiaceae	*Cola gigantea* var. glabrescens
38	Sterculiaceae	*Triplochiton scleroxylon*
39	Rubiaceae	*Mitragyna stipulosa*

This is shown by particular marks inside the RFs (e.g. on specific tree species, on the soil, on stones, etc.) that indicates restricted access to the water and RF. Ceperley et al. (2010) have shown that sacred pools or points in the river along the Ouèmé and Okpara Rivers in Central Benin with their adjacent RFs are governed by rules that influence conservation practices, including bans on fish poisoning, over-fishing and pollution and discouragement of cattle grazing and cultivation in their vicinity. RFs structure and diversity were examined in three sites adjacent to sacred pools as well as in riparian areas adjacent to land uses in the region. Diversity in riparian forest tree species was highest in areas adjacent to sacred forests than in areas adjacent to village uses. Concurrent interviews with elders and hunting chiefs revealed the significant cultural importance of sacred pools and riparian resources. This would be crucial to flood control and biodiversity conservation in central Benin.

5. Issues and Challenges to Sustainable Management of RFs

5.1. Impacts of Climate Change on RFs

In sudanian savanah in North Benin and Togo, natural mortality of the trees in RFs is very high (21.2%). It affected more trees of small diameter. Adjonou et al. (2009) indicated that the cause of this high mortality was the effects of recent climate change in the region by temperature increase and the decrease of rainfall. The consequences of these changes were the dry up of the climate and the reduction of water and nutrients availability, minimizing the physiological functioning of the small diameter trees.These modifications also appear through the increase of fire risks and the harshness of the bush fire. Indeed, the mortality in vegetation could be explained by many natural, pathological, anthropic and climatic causes. Mortalities due to events such as natural degeneration or pathologic infections were easily detectable on the field. But an important trees' mortality on a large surface and during a long period could be under internal conditions (health of the trees) or external conditions (ecology and climate) (Allen and Breshears 1998). Many studies addressed that forest ecosystems and woody species in Africa were submitted to the incidence of temperature increase in dry period because of climate change (Darkoh 2003, Sarr and Roy-Macauley 2007). Other specific studies showed that consecutive droughts between 1970 and 1980 in Africa have directly increased the mortality of woody plants species from fragile ecosystems.Indirectely, the extension of heat period and the drying of areas lead to the activation of physiological stress inside the trees that could reduce his immune system against pathologic attacks (Wardell et al. 2003).These situations are currently recurrent in the RFs in sudanian zone (Photo 2).

Photo 2. Tree mortality in a RF of Oti-Keran National Park in sudanian zone of Togo.

Another natural pressure is the phenomenon of ground sinking called *Donga* on tree species structure in RFs (Photo 3). This phenomenon was studied by Fandohan (2011) in north Benin. The author mentioned that there are signicant changes and a noticeable decline in number of adult and juvenile *Tamarindus indica*, a multipurpose tree, in the site subjected to ground sinking. *Dongas* critically limit regeneration strategies other than root sucking i.e. recruit issued from dispersed seeds. Findings would suggest that the resultant effect of *Dongas* may lead to a decline in the species' populations (Fandohan, 2011).

Photo 3. Ground sinking or *Donga* phenomenon in a RF of Oti-Keran National Park in North Togo.

5.2. Threats to Riparian Forests Biodiversity through Agriculture, Mining and Pollution

The consequences of human activities on water quality and quantity, plants composition, growth and regeneration, physiognomy, sediments deposit, erosion, flood, fluvial processes, density and variety of animals, etc. are always tremendous and dramatic (Kellman and Meave 1997, Levine and Stromberg 2001).

In the DG, the rapid changes in land use modify the structure, composition, hence the functions of RFs, which provide valuable and scarce plants and animals, fertile soil for cultivation, give an opportunity for irrigation, and water for herds of cattle in transhumance. Their rich biological resources are disappearing before they can be inventoried and assessed. Native RFs have been severely degraded, through dam construction, water diversions, draining of waterways, lowering of water tables, river channelisation, cattle ranching, agricultural development, recreational development, urbanisation on lands adjacent to RFs, vegetation fire, forest clearance, selective tree cutting, invasion of exotic plants, poaching, water poisoning (Roods et al. 1995, Busch and Smith 1995, Molles et al 1998, Natta et al. 2002).

Photo 4. Farming in the RF along Koumongou River.

Anarchic forest logging by local people remains the first menace for RFs. Species like *Mitragyna stipulosa*, *Aubrevillea kerstingii, Pterocarpus mildbraedii*, *Afzelia africana*, and *Khaya anthotheca* were endangered in RFs because of traditional timber logging and firewood production (Fontodji et al. 2009, Gentry 1992).

The new forms of landuse consisting to clear RFs in favour of gardening or cropping are the permanent dangers (Photo 4). The most dangerous forms of clearing are the cultivation in the source of the rivers. These practices reduce considerably the size of the RFs and lead to their dry up.

The introduction of exotic plant species with rapid growth in the RFs became a serious threat; the artificial vegetation replaces progressively the natural vegetation. Invasive species recorded in the RFs are *Azadirachta indica*, *Cedrela odorata*, *Delonix regia* and *Tectona grandis*.

Artisanal mining is also a considerable threat. Indeed gold and diamond deposits were present under some RFs or were extracted in the bed of the river, with modern means which require the clearing of the RFs.

6. Restoration Strategies of the RFs in the DG

6.1. Ecological Indicateurs as Sustainable Management Tools

Benin and Togo Republics have signed and ratified the great conventions relative to sustainable development (Convention on Climate Change, Biodiversity, desertification, etc). This commitment emphasised to both countries the responsibility for conserving and exploiting its own biological diversity on a sustainable basis, and to integrate their management in the sectorial and intersectorial plans, programs and policies. Therefore many programs of reforestation and forests restoration are under way in these countries. Forest departments and NGO involved in the implementation of those programs take into account the importance of ecological, economic and social indicators in every forest type sustainable management.

For instance, as ecological indicators, this study shows that RFs are refuges of biodiversity. Concerning economical indicators, a liste of 40 rare and threatened species of timber and many other economical value non-wood forest products species had been recorded inside RFs. This results allow to conclude that RFs in the DG fulfiled to the criteria and indicators (CandI) of sustainable forest magement, including the conservation of sensitive zones (Principle 3, Criterion 3.3, Indicators 3.3.2 and 3.3.4) (Wijewardena 1998, OAB-OIBT 2003, Prabbu et al. 2005). This includes also the protection of water resources (Principle 3, Criterion 3.5, Indicators 3.5.1 and 3.5.2). CandI give a particular importance to natural regeneration and the use of local species for reforestation or for agroforestry (Principle 3, Criterion 3.4, Indicators 3.4.1 and 3.4.2). Many species listed in RFs (*Amphimas pterocarpoides, Aubrevillea kerstingii, Bombax buenopozense, Canarium schweinfurthii, Ceiba pentandra, Distemonanthus benthamianus, Entandrophragma cylindricum, Khaya spp., Milicia excelsa, Mitragyna stipulosa, Piptadeniastrum africanum, Pterocarpus mildbraedii, Pterygota macrophylla, Pycnanthus angolensis, Sterculia oblonga, Symphonia globulifera, Terminalia superba* et *Triplochiton scleroxylon,* etc), could be used in reforestation in both countries. This study showed also that about twenty species of wild yams represent an important genetic base for yam crops improvement (Hladik and Dounia, 1996). RFs offer many marketable non-wood forest products (e.g. *Dialium guineense*, edible muschrooms, honey, fish, meat, shellfish) and important touristic opportunities such as waterfall and footpaths.

Photo 5. Clothes' washing in the watercourse close to a village.

Concerning social CandI favourable to the sustainable management of RFs, some opportunities exist. Indeed, local populations had before the habit to conserve a fringe of forest along the rivers in their farms. Apart from undergrowth crops, it is unusual to farming close to bed's river. In fact, RFs were considered as common property (for a family, lineage or community) but also a private property because nobody could exploit the RFs of another. Some cases of community management of RFs were noticed on the field where members of the same family share the resources. That's how forest products harvested in RFs could be sold settling problem of the family. Many rivers and watercourses were managed for water supply in the village (Photo 5). Some village considered rivers as sacred site and the river's spirit were located in the forest along the bed. The taboos associated to these practices were the prohibition to clear the RFs for farming or timber logging. Most of these taboos aimed to prevent the pollution by pesticides use.

6.2. Policies and Rules for Riparian Forests Management

In order to protect biodiversity in natural landscapes and regulate the utilisation of biological resources, each government (Benin and Togo) decreed a new forest law (n°93-009 of July 1993 for Benin and n°2008-009 of 19 june 2008 for Togo). The uniqueness, vulnerability and diversity of RFs are the most important criteria to protect them. The forest laws recognize RFs as refuge ecosystems for plants and wildlife of many kinds. For example, clearance of wood and shrubs is not allowed within 25 m at both sides of any waterway (article 28 in Benin and article 56 in Togo). Despite the strictness of these conservation policies against clearance of RFs, cutting of protected trees, hunting of protected and rare animal species, in reality uncontrolled, unplanned, and mostly illegal utilisations go on, especially in non-protected areas.

RFs width varies from river to river, and along a single river system. This has several vital functional implications. The legal limit of 25 m to be protected at both sides of any waterway was not based on scientific data. Natta et al. (2002) suggested a limit of 100 m to be protected based on field observations. Further vegetation relevés across RFs and adjacent vegetations in Benin give insight to the issue of optimal width to be protected. A main research question remains; how wide should riparian forest corridors be to fulfil their vital ecological functions?

A prudent approach for biodiversity conservation is to preserve not only the full range of river front, interior and edges species of RFs (i.e. at least 57 m from the river bed, see Natta 2003), but also parts of the adjacent upland forest or savanna as buffer zone. Moreover dense and continuous canopy at the riverside can expend to more than 100 m. This could guide legislators to update the optimal threshold distance to be protected at each riverside at 100 m (*i.e.* four times the actual distance). In similar ecosystems, overseas studies have shown the importance of the issue of optimal distance to be protected at both waterway sides. In South East Brazil, the optimal width that allows RFs to fulfil their multifunctional roles (e.g. sediment retention and water quality improvement) was found to be 52 m, which was actually wider and more efficient than the official legal recommendation of 30 m (Sparovek et al. 2002). In South East Australia, Dignan and Bren (2003) found that the fixed minimum non-disturbance width of 20 m along permanent streams does not provide maintenance of the light regime that could protect RFs, instead they suggest forested buffers 70 to 100 m wide.

6.3. Restoration Strategies

Effective management strategies must be devised to preserve and restore fluvial systems, and riparian woodlands and animals while relying on accurate understanding of the structure and dynamics of RFs communities. However, simply protecting RFs in a buffer zone may not be adequate to ensure their existence in the long term (Hibbs and Bower 2001). Instead the management of RFs must be a component (*i.e.* subset) of watershed or landscape management. On the other hand, awareness has to be raised with all stakeholders on RFs as unique physical and natural systems in their own right, and warranting special management and protection.

Research in Australia has shown that cost-effective conservation of biodiversity requires that maximum biodiversity should be protected in a minimum area (Williams et al. 1991). The current degraded status of many RFs in savanna regions represents the cumulative and persistent impacts of unplanned and uncontrolled land uses and poor management of RFs themselves but also of adjacent upland vegetations. Strategies that reflect a spectrum of goals linked to improving the ecological functions and sustainability of riparian ecosystems (i.e. natural integrity and diversity of the system) are needed. For example, strategies that focus on returning the hydrologic regime to a more natural state have the greatest potential for restoring riparian vegetation, and forests in particular that have evolved with and adapted to the patterns of changing flows in waterway environments. As pointed out by Cordes *et al.* (1997), to maintain existing and restore riparian woodland and to ensure their long-term regeneration, it has become important to understand in some detail both the contemporary and historical relationships between riparian vegetation and hydrological and geomorphological factors.

Also, management practices to maintain RFs biodiversity need to be tailored to the conditions of each particular area (Hancock et al. 1996), in particular one must take into consideration the types of disturbances that typically affect these forests. Therefore, activities aiming at protecting and rehabilitating RFs in their watershed should be implemented at the national, regional, and local levels, while taking into account ecological benefits (i.e. physical variables) as well as social costs represented by the loss of arable land (Xiang 1993).

Because RFs perform a more than proportionate number of critical biological and physical functions on a unit area basis, their restoration when degraded, and protection when undisturbed will have a major influence on the local and regional environment. Meanwhile, deciding on priorities between rehabilitation and preservation may be difficult. Following Geraghty and Vollebergh (1993), it is more cost effective initially to maintain systems that are in reasonable condition and build from them, rather than at first attempting difficult rehabilitation tasks of much degraded river sections.

Capitalising from the issues and challenges regarding the participatory management of RFs, keys assumptions stated bellow as follow:

- Landscape patterns that promote connectivity for species, communities, and ecological processes are key elements for nature conservation in environments modified by human impacts (Bennett 1999). This is particularly true in savannah landscapes where remnant RFs play the role of forest ecosystems;
- RFs can be considered biodiversity 'hotspots', particularly in savannah regions, and deserve special care (Natta 2003);
- Municipalities authorities must develop/implement participatory landscape management plans, based on effective zoning and ecological potential of each region;
- The multipurpose and participatory management of RFs should be promoted in protected, as well as in non protected areas;
- The integrated management of RFs must be a subset of good watershed management plan, where all vegetations and landscape features play a role, and that takes into account the conflicting needs of all stakeholders;
- When implementing landscape management plans, managers of RFs biodiversity have to take decisions between (a) preservation of mature or less disturbed stands, (b) ecological restoration of degraded RF stands, and (c) rehabilitation of converted RFs or very degraded river sections;
- There is a great need to base restoration, rehabilitation, and preservation strategies on accurate scientific knowledge about the structure and dynamics of riparian systems;
- The preservation of intact or little disturbed RF stands is vital because they represent valuable reference sites for understanding the goals and efficacy of restoration efforts. Also, they are important sources of genetic material for the reintroduction of native species into degraded areas;
- The restoration of RFs should refer firstly to the process of removing the degrading factor or reducing its negative effects, and secondly repairing the condition and functioning of degraded riparian vegetation. Ecological restoration of degraded RFs, in which biodiversity is greatly decreased, has the stated goal to enhance the value of the ecosystem, moderate degrading influences and regain pre-disturbance characteristics (NRC

2002). This approach may include planting native trees, facilitating development of certain species, preserving and rehabilitating threatened and endemic species. The problem of isolated populations could be partially solved by reforesting corridors between RFs patches and belts along watercourses;

- The rehabilitation of converted RFs or much degraded waterways sections may not be cost-effective as RFs biodiversity conservation strategy. Meanwhile this option can be adopted for recreational, research, and eco-tourism purposes in urbanised landscapes;
- Qualitative and quantitative knowledge of hydrology, fluvial ecology, species and ecosystems diversity, natural variability, disturbance regimes, soils and landforms, are therefore required for successful preservation, restoration and rehabilitation of RFs in protected and non protected areas;
- Legislation and policies for the protection or rehabilitation of RFs, must take into account the economic and social costs as well as the environmental benefits at various scales. Before improving the actual forest law in both countries, there is an urgent requirement to increase enforcement of existing legislation, which is often unknown or ignored by many stakeholders. So, as suggested by certains authors (e.g. de Lima and Gascon, 1999; Natta, 2007, etc.), proactive educational initiatives are direly needed and should be coupled with legislation enforcement to reduce deforestation of riverbanks and maintain corridors between ecosystems in populated regions;
- Foresters, rural extension workers, and NGOs dealing with natural resource protection must promote local people's rights and duties specified by the forest law. Effective conservation can best be achieved if the local population is either collaborating or in control. Initiatives have to be taken to rise awareness concerning their benefits in protecting RFs (e.g. watershed, water sources and species protection, landscape beauty, healthy environment, sustainable harvest of NTFP);
- In Participatory Management Plans (PMP) for forest reserves, RFs are to be left uncut, clearance of vegetation is not allowed within 25 m at waterways sides, and rare species outside the RFs will not be cut either;
- In rural lands and urban zones, selective tree cutting, cattle ranching, irrigation, water pumping and poaching in riparian zones should be regulated. Multilayered RFs are systematically removed and replaced by farms, open fields, or savannas of much less ecological value. Apart from sacred stands and water sources where traditional rules are respected, nothing is done to protect RFs;

- The preservation of rare and endangered species should be coupled with the protection of their habitats. So, the distribution pattern of the endemic primate sub-species (i.e. *Cercopithecus erythrogaster erythrogaster*) of Benin, suggest protecting all the RFs in the Ouémé valley, which are among its most important habitats (Sinsin et al. 2002). Likewise, RFs along Yarpao stream (Natitingou district) and many others in the Atacora hills should have higher legal protection priority for their great plant diversity and for protecting the ornamental *Thunbergia atacorensis*;
- From the issues, challenges and strategies dedicated to RFs biodiversity conservation stated in the sections above, simply protecting RFs in a buffer zone may not be adequate to ensure their existence in the long term. Instead the integrated management of RFs must be a component (i.e. subset) of good watershed or landscape management that takes into account the conflicting needs of all stakeholders. Therefore within the framework of democratisation and decentralisation, local authorities must develop Education, Information and Communication programs that promote awareness on biodiversity protection, and landscape beauty valorisation through recreation, research and eco-tourism.

Conclusion

The complexity of the RFs is the origin of their exceptionnal biological diversity shelting. About 1,200 plant species were recorded in the RFs in the DG, i.e. more than 40% of the flora in Benin and Togo. So this study confirms that RFs constitute the hotspot and refuges for biodiversity in the DG. They support a high level of richness. Therefore vis-a-vis the retreat of the forest vegetation everywhere in Benin and Togo, the studied riparian forests have today particularly important value. The presence of the endemic, rare and unknown species confers to them zone of high value for conservation (ZHVC). For instance, facing to forest degradation everywhere in both countries, RFs appear nowdays like refuges of the local biological diversity.

In line with the Convention on Biological Diversity which emphasised the need for every country to take responsibility for conserving its own biodiversity, a prudent approach to biodiversity conservation necessitates the preservation of all potential rich or endangered ecosystems. The maintenance of RFs is recognised as being vital to the integrity of waterways and biodiversity conservation. Given that RFs can support high levels of biodiversity, and have played an important role during periods of climate change over geological time, their conservation in savannah ecosystems is of utmost importance with the threat of increase in human disturbance and change in climate.

As significant as climate changes are likely to be, land- and water-use changes have had and will continue to have the greatest effects on riparian vegetation, so efforts toward the conservation of RFs will increasingly become a priority in land use planning. Therefore, more qualitative and quantitative assessments of the effects of land and water management on RFs have to be undertaken. The main variables to be considered should be related to physical, biological and anthropogenic factors.

Despite the forest laws, the protection of RFs is very difficult. Different stakeholders dealing with biodiversity conservation should take initiatives for the preservation of the remaining rich and diverse RFs especially in hitherto non-protected areas. Integrated management of RFs that optimise their values as habitat for native plants and animals requires planning and acting, with all stakeholders, at both site-specific and watershed/landscape levels. Through improved and long-term scientific, educational and recreational programs, the ecological importance and intrinsic human values associated with RFs may be better balanced against the competing wants of unplanned urbanisation, short-term local inhabitants' demands, and unknown motivations of other stakeholders.

REFERENCES

Adjanohoun, E., Ahyi, M.R.A., Ake Assi, L., Akpagana, K., Chibon, P., El-Hadji, A., Eymê, J., Garba, M., Gassita, J.N., Gbeassor, M., Goudote, E., Guinko, S., Hodouto, K.K., Houngnon, P., Keita, A., Keoula, Y., Kluga-Ocloo, W.P., Lo, I., Siamevi, K.M., Taffame, K.K., 1986. *Contribution aux études ethnobotaniques et floristiques au Togo, ACCT,* Paris, 671 p.

Adjanohoun, E.J., Adjakidjè, V., Ahyi, M.R.A., Ake Assi, L., Akoègninou, A., d'Almeida, J., Apovo, F., Bouke, F.K., Chadare, M., Cusset, G., Dramane, K., Eyme, J., Gassita, J.-N., Gbaguidi, N., Goudote, E., Guinko, S., Houngnon, P., Issa, L.O., Keita, A., Kiniffo, H.V., Kone-Bamba, D., Musampa Nseyya, A., Saadou, M., Sodogandji, Th., De Souza, S., Tchabi, A., Zinsou Dossa, C., and Zohoun, Th.,1989. *Contribution aux études ethnobotaniques et floristiques en République Populaire du Bénin. Agence de Coopération culturelle et technique, Paris,* 895 pp.

Adjonou, K., Bellefontaine, R., Kokou, K., 2009. Les forêts claires du parc national Oti-Kéran au Nord-Togo : structure, dynamique et impacts des modifications climatiques récentes. *Sécheresse 2009* ; 1E (5) : 1-10.

Adjossou, K., 2004. *Diversité floristique des forêts riveraines de la zone écologique IV du Togo*. Master thesis in Biologie de Développement, Option Biologie Végétale Appliquée, University of Lomé. 75 pp.

Adomou, A.C., 2011. Phytogeography of Benin, 14-20. In: Neuenschwander, P., Sinsin, B., Georgen, G. (eds). Protection de la Nature en Afrique de l'Ouest : *Une Liste Rouge pour le Bénin*. International Institute of Tropical Agriculture, Ibadan, Nigéria. 365 p.

Akoègninou, A., 1984. Contribution à l'étude botanique des îlots de forêts denses humides semi-décidues en Rébublique Populaire du Bénin. *Thèse de troisième cycle,* Université de Bordeaux 3, Bordeaux, France. 181p.

Akpagana, K., 1989. Recherches sur les forêts denses humides du Togo. *Thesis de Doctorat*, University of Bordeaux III, Bordeaux, France. 181 p.

Allen, C.G., Breshears, D.D., 1998. Drought-induced shift of a forest-woodland ecotone: rapid landscape response to climate variation. *Proceedings of the National Academy of Sciences;* 95 : 14839-42.

Baker, W.L., 1990. Species richness of Colorado riparian vegetation. *Journal Vegetation Science* 1:119-124

Bendix, J., Hupp, C.R., 2000. Hydrological and geomorphological impacts on riparian plant communities. *Hydrological Processes* 14:2977-2990

Bennett, A.F., 1999. Linkages in the landscape: the role of corridors and connectivity in wildlife conservation. IUCN, Gland, Switzerland and Cambridge, UK. 254p.

Bornette, G., Amoros, C., 1996. Disturbance regimes and vegetation dynamics : role of floods in riverine wetlands. *Journal of Vegetation Science* 7:615-622

Bornette, G., Amoros, C., Rostan, J.C., 1996. River incision and decennial vegetation dynamics in cut-off channels. *Aquatic Sciences* 58(1):31-51

Bornette, G., Amoros, C., Piégay, H., Tachet, J., Hein, T., 1998. Ecological complexity of wetlands within a river landscape. *Biological Conservation* 85:35-45

Bowman, D.M.J.S., McDonough, L., 1991. Tree species distribution across a seasonally flooded elevation gradient in the Australian monsoon tropics. *Journal of Biogeography* 18:203-212

Brunel, J.F., Scholz, H., Hiepko, P., 1984. *Flore analytique du TOGO. Phanérogames.*GTZ, Eschorn, 751 pp

Busch, D.E. Smith, S.D. 1995. Mechanisms associated with decline of woody species in riparian ecosystems of the southwestern US. *Ecological Monographs* 65(3): 347-370.

Ceperley N., Montagnini F., Natta, A. 2010. Significance of sacred sites for riparian forest conservation in Central Benin. *Bois et Forêts des Tropiques*, Biodiversité forêt galerie et sites sacrés. 303 (1) 5-23.

Da Costa Filho, L.V., Nanni, M.R., Campos, J.B., 2006. Floristic and phytosociological description of a riparian forest and the relationship with the edaphic environment in Caiuá ecological station - Paraná – Brazil. *Brazilian archives of Biology and Technology* 49:785-798.

Darkoh, M.B.K., 2003. Regional perspectives on agriculture and biodiversity in the drylands of Africa. *J Arid Environ* 54 : 261-79.

Décamps, H., Naiman, R.J., 1989. L'Ecologie des fleuves. *In* : *Revue Recherche,* n° 208.

Détienne, P., 1991. Clé de reconnaissance par les feuilles des légumineuses des forêts du Cameroun, Congo, Gabon et de la R.C.A. *Revue Bois et forêts des Tropiques*, n° 230, 4e trimestre.

Dignan, P., Bren, L., 2003. A study of logging on the understorey light environment in riparian buffer strips in a south-east Australian forest. *Forest Ecology and Management* 172: 161-172.

Cordes, L.D., Hughes, F.M.R., Getty, M., 1997. Factors affecting the regeneration and distribution of riparian woodlands along a northern prairie river: the red Deer River Alberta, Canada. *Journal of Biogeography*24: 675-695.

Lima (de), M.G., Gascon, C., 1999. The conservation value of linear forest remnants in central Amazonia. *Biological Conservation* 91: 241-247.

Ettema, C.H., Lowrance, R., Coleman, D.C., 1999. Riparian soil response to surface nitrogen input: the indicator potential of free-living soil nematode populations. *Soil Biology and Biochemistry* 31: 1625-1638.

Ern, H., 1979. *Die Vegetation Togos*. Gliederung, Gefährdung, Erhaltung. Willdenowia 9: 295-312.

Fandohan, B., 2011. *Conservation biology of Tamarindus indica* (Fabaceae) in Benin, West Africa.PhD Thesis. University of Abomey-Calavi, Benin. 227p.

FAO, 2003. Proceedings of second international workshop on participatory forestry in Africa. 18-22 February 2002 Arusha Tanzania.

Ferreira, L.V., 1997. Effects of the duration of flooding on species richness and floristic composition in three hectares in the Jau National Park in floodplain forests in central Amazonia. *Biodiversity and Conservation* 6:1353-1363

Fetherston, K.L., R.J. Naiman, Billy, R.E., 1995. Large woody debris, physical process, and riparian forest development in montane river networks of the Pacific Northwest. *Geomorphology* 13: 133-144.

Fontodji, K.J., Mawussi, G., Nuto Y., Kokou, K., 2009. Effects of charcoal production on soil biodiversity and soil physical and chemical properties in

Togo, West Africa. *International Journal of Biolology and Chemistry Sciences* Vol. 3, No. 5, pp. 870-879

Forman, R.T.T., Godron, M., 1986. *Landscape ecology*. John Wiley and Sons. New York. 619p.

Gentry, A.H., 1991. The distribution and evolution of climbing plants. In: Putz, F.E. and Mooney H.A. (eds.). *The biology of vines*, pp. 3-47. Cambridge University Press.

Geraghty, P., Vollebergh, P., 1993. Riparian zone management in Victoria: Policy and practice. In: Bunn, S.E., B.J. Pusey and P. Price (eds.), *Ecology and Management of Riparian Zones in Australia*. LWRRDC Occasional Paper Series No: 05/93.

Giese, L.A.B., Aust, W.N., Kolka, R.K., Trettin, C.C., 2003. Biomass and carbon pools of disturbed riparian forests.*Forest Ecology and Management* 6198: 1-16.

Goebel, P.C., Palik, B.J.; Pregitzer, K.S., 2003. Plant diversity contributions of riparian areas in watersheds of the Northern Lake States, USA. *Ecological Applications* 13:1595-1609

Gregory, S.V., Swanson, F.J., Mskee, W.A., Cummins, K.W., 1991. An ecosystem perspective of riparian zones: focus on links between land and water. *Bioscience* 41:540-551.

Grubb, P., Lernould J.M., Oates, J.F., 1999. Validation of *Cercopithecus erythrogaster pococki* as the name for Nigerian white-throated guenon. *Mammalia* 63(3): 389-392.

Guillaumet, J.L, Adjanohoun, E., 1971. La végétation de la Côte d'Ivoire. In *: Le milieu naturel de la Côte d'Ivoire*. Mémoire ORSTOM n° 50 : 156-263.

Harper, D., C. Smith, P., Barham, Howell, R., 1994. The ecological basis for the management of the natural river environment. In: Harper, D. and A.J.D. Ferguson (eds.), *The ecological basis for river management*. Wileym Chichester. pp 219-238.

Hall, J.B., Swaine, M.D., 1981. Distribution and ecology of vascular plants in a rain forest vegetation. *In Ghana.Geobotany* 1, 383 p.

Hancock, C.N., Ladd, P.G., Froend, R.H., 1996. Biodiversity and management of riparian vegetation in Western Australia. *Forest Ecology and Management* 85: 239-250.

Heywood, V.H., Davis, S.D., 1994. *Introduction*. In : Davis, S.D.; Heywood, V.H.; Hamilton, A.C. (eds) Centres of Diversity. *A guide and strategy for their conservation*. IUCN, WWF, ODA, EU, UK, 1-38

Hladik, A., Dounias, E., 1996. «Les ignames spontanées des forêts denses africaines. Plante à tubercule comestible», *in* : Hladik C.M., Hladik A.,

PagezyH., Linare O.L., Koppert G.J.A. et Froment A., *L'alimentation en forêts tropicales. Interaction bioculturelle et perspective de dévéloppement.* Unesco, Paris, pp. 275-294.

Hibbs, D.E., Bower, A.L., 2001. Riparian forests in the Oregon Coast Range.*Forest Ecology and Management*154: 201-213.

IUCN (Sayer, J.A., Harcourt, C.S., Collins, N.M. (eds.)), 1992.*The conservation atlas of tropical forest Africa.* Macmillan Publishers Ltd.

Jones, A.P.D., 1945. Notes on terms for uses in vegetation description in southern Nigeria. *Farm and Forest* 6: 130-136.

Jones, E.W., 1955. Ecological studies on the rain forest of southern Nigeria.IV.The plateau forest of Okumu Forest reserve. Part I. The environment, the vegetation types of the forest and the distribution os species. *Journal of Ecology* 43: 564-594.

Jongkind, C.C.H., 2004. *Checklist of Upper Guinea forest species* In: Poorter, L.; Bongers, F., N'Kouamé, F., Hawthorne, W.D. (eds) Biodiversity of West African Forests. An ecological atlas of woody plant species. CABI Publishing, Oxford, UK, 447-474.

Keay, R.W.J., 1949. An example of Sudan zone vegetation in Nigeria.*Journal of Ecology* 37: 355-364.

Keay, R.W.J., 1959. Derived savanna – derived from what? *Bulletin IFAN série A* 21: 427-438.

Keay, R.W.J., 1960. An example of Northern Guinean Zone vegetation in Nigeria.*Nigeria Forestry Information Bulletin* 1: 1-46.

Keay, R.W.J., Onochie, C.F.A., Standfield, D.P., 1964. Nigerian trees. *Federal Department of Forest Research, Ibadan, Nigeria* Vol 1 (334p), Vol 2 (395p).

Kellman, M., Meave, J., 1997. Fire in the tropical gallery forests of Belize. *Journal of Biogeography* 24 (1): 23-34.

Kolongo, T.S.D., Decocq, G., Adou Yao, C.Y., Blom, E.C., van Rompaey, S.A.R., 2006. Plant species diversity in the southern part of the Taï National Park (Côte d'Ivoire). *Biodiversity and Conservation* 15:2123-2142

Kokou K., Caballé, G., 2000. Les îlots forestiers de la plaine côtière togolaise. *Bois For. Trop.* 263(1): 39-51.

Kokou, K., Adjossou, K., Kokutse, A. D., 2008. Considering sacred and riverside forests in criteria and indicators of forest management in low wood producing countries: The case of Togo. *Ecological indicators* 8, pp 158-169

Kouamé, F.N., Kouadio, K.E., Kouassi, K., Poorter, L., 2004. Floristic diversity of the closed forests in Côte d'Ivoire. In: Poorter, L., Bongers, F., N'Kouamé, F., Hawthorne, W.D. (eds) Biodiversity of West African Forests. *An ecological atlas of woody plant species.* CABI Publishing, Oxford, UK, 53-60

Langenberger, G., Konrad, M., Sauerborn, J., 2006. Vascular plant species inventory of a Philippine lowland rainforest and its conservation value. *Biodiversity and Conservation*15:1271-1301

Letouzey, R., 1982. *Manuel de Botanique forestière*. Afrique Tropicale. Centre technique forestier tropical.Tome1, 131 – 140 p.

Levine, C.M., Stromberg, J.C., 2001. Effects of flooding on native and exotic plant seedlings: implications for restoring south-western riparian forests by manipulating water and sediment flows. *Journal of Arid Environments* 49: 111-131.

Malm Renofalt, B., Merritt, D.M., Nilsson, C., 2007. Connecting variation in vegetation and stream flow: the role of geomorphic context in vegetation response to large floods along boreal rivers. *Journal of Applied Ecology* 44:147-157

Meave, J., Kellman, M., 1994. Maintenance of rain forest diversity in riparian forests of tropical savannas: implications for species conservation during Pleistocene drought. *Journal of Biogeography* 21:121-135

Meave, J.; Kellman, M.; MacDougall, A., Rosales, **J., 1991. Riparian habitats as tropical forest refugia.** *Global Ecology and Biogeography Letters* 1:69-76

Medley, M.E., 1992. Patterns of forest diversity along the Tana River, Kenya. *Journal of Tropical Ecology* 8:358-371

Molles, Jr.M.C., Crawford, C.S., Ellis, L.M., Valett, H.M.,Dahm, C.N., 1998. Managed floods: restoration of riparian forest ecosystem structure and function along the Middle Rio Grande. *BioScience* 48: 749-756.

National Research Council, (NRC), 2002. *Riparian areas: Functions and strategies for management*. Summary. WSTB, CRZFSM. National Academy Press, Washington. 22p.

Natta, A.K., Sinsin, B., van der Maesen, L.J.G., 2002. Riparian forests, a unique but endangered ecosystem in Benin. Notulae Florae Beninensis 4. *Botanische Jahrbucher* 124 (1): 55-69.

Natta, A.K., Sinsin, B. van der Maesen, L.J.G., 2003. A phytosociological study of Riparian forests in Benin (West Africa). *Belgian Journal of Botany* 136 (2): 109-128.

Natta, A.K., 2003. *Ecological assessment of riparian forests in Benin: phytodiversity, phytosociology, and spatial distribution of tree species.* Doctoral thesis, University of Wageningen, Netherland, 215 p.

Natta, A.K., 2007. Restoration of riparian forests ecological diversity in landscape management plans: issues and lessons from Benin. pp 21 - 34. In: Mayaka T.B., de Iongh H. and Sinsin B. (eds.) (2007*). Ecological Restoration of African Savanna Ecosystems*. Proceedings of the third Regional Network for the Synergy Between the Conservation on Biological Diversity and the Convention to Combat Desertification in West and Central Africa (RNSCC) International Seminar, 6 February, Cotonou, Benin. CEDC - Cameroon /CML, Leiden University, the Netherlands.156 p.

Nilsson, C., Grelsson, G., Johansson, M., Sperens, U., 1989. Patterns of plant species richness along riverbanks. *Ecology* 70:77-84.

OAB-OIBT, 2003. Principes, critères et indicateurs OAB-OIBT de la gestion durable des forêts tropicales naturelles d'Afrique. *Série Développement de politiques OIBT n°* 14, 26 p.

Oates, J.F., 1996. Survey of *Cercopithecus erythrogaster* populations in the Dahomey Gap. *African Primates* 2(1): 9-11.

Pither, R., Kellman, M., 2002. Tree species diversity in small, tropical riparian forest fragments in Belize, Central America.*Biodiversity and Conservation* 11:1623-1636

Prabbu, R., Golfer, C., Shepherd, G., 2005. Critères et indicateurs d'une gestion forestière durable : nouveaux résultats des recherches du CIFOR au niveau de l'unité de gestion forestière. *ODI, Document RDFN numéro 23a*, pp 1-31.

Roods, S.B., J.M. Mahoney, D.E. Reid, L. Zilm, 1995. Instream flows and the decline of riparian cottonwoods along the St. Mary River, Alberta.*Canadian Journal of Botany* 73:1250-1260.

Sarr, B., Roy-Macauley, H., 2007. Tendances actuelles et futures du climat et impacts potentiels sur les ressources forestières. Communication à l'Atelier sous-régional sur les *«Leçons apprises et les perspectives de la gestion durable des forêts au Sahel* », Bamako (Mali).

Ssegawa, P., Nkuutu, D.N., 2006. Diversity of vascular plants on Ssese islands in Lake Victoria, central Uganda. *African Journal of Ecology* 44:22-29

Sinsin, B., Nobimé, G., Téhou, A., Bekhuis, P. Tchibozo, S., 2002. Past and present distribution of red-bellied monkey *Cercopithecus erythrogaster erythrogaster* in Benin. *Folia primatologia* 73: 116-123.

Sokpon, N., 1995. Recherches écologiques sur la forêt dense semi-décidue de Pobè au sud-est du Bénin : groupements végétaux, structure, régénération naturelle et chute de la litière. *Thèse Doct. Univ. Libre de Bruxelles*, 293 p.

Sokpon, N., Sinadouwirou, T., Gbaguidi, F. and Biaou, H., 2001. Aperçu sur les forêts édaphiques hygrophiles du Bénin. *Belgian Journal of Botany,* 134 (1), 79-93.

Sparovek, G., Ranieri, S.B.L., Gassner, A., de Maria, I.C. Schnug, E.E., dos Santos R.F., Joubert A., 2002. A conceptual framework for the definition of the optimal width of riparian forests.*Agriculture, Ecosystems and Environment* 90: 169-175.

Tabacchi, E., Planty-Tabacchi, A., Decamps, O., 1990. Continuity and discontinuity of the riparian vegetation along a fluvial corridor. *Landscape Ecology* 5:9-20

Tabacchi, E., Planty-Tabacchi, A., 1996. Landscape structure and diversity in riparian plant communities: a longitudinal comparative study. *Research and Management* 12:367-390.

Tchouto, M.G.P., De Boer, W.F.; De Wilde, J.J.F.E., van der Maesen, L.J.G. 2006. Diversity patterns in the flora of the Campo-Ma'an rain forest, Cameroon: do tree species tell it all? *Biodiversity and Conservation* 15:1353-1374

Wardell, D.A., Reenberg, A., Tettrup, C., 2003. Historical footprints in contemporary land use systems: forest cover changes in savannah woodlands in the Sudano-Sahelian zone. *Global Environmental Change-Human and Policy Dimensions* 13 : 235-54.

White, F., 1983. The vegetation of Africa, a descriptive memoir to accompany the UNESCO/AETFAT/UNSO vegetation map of Africa. UNESCO, *Natural Resources Research* 20:1-356

Wijewardena, D., 1998. «Critères et indicateurs de gestion durable» (Criteria and indicators for sustainable forest management). *Bulletin de l'OIBT sur les forêts tropicales,* Vol 8 (3), 4-6.

Williams, P.H., Humphries, C.J., Vane-Wright, R.I., 1991. Measuring biodiversity: Taxonomic relatedness for conservation priorities. *Australian Systematic Botany* 4: 665-79.

Xiang, W.N., 1993. Application of a GIS-based stream buffer generation model to environmental policy evaluation. *Environment Management* 17: 817-827.

In: Riparian Zones
Editor: A. Kerem et al.
ISBN 978-1-61942-677-1

Chapter 3

Partial Geomorphic and Floristic Recovery of an Incised Channel over the Last Twenty Years on Dairy Arm, Hunter Valley, Australia

Anita Chalmers[1], Wayne D. Erskine[1,2] and Fiona Erskine[1]
[1] School of Environmental and Life Sciences, The University of Newcastle, Ourimbah Campus, PO Box 127, Ourimbah NSW 2258 Australia.
[2] Environmental Research Institute of the Supervising Scientist, GPO Box 461, Darwin NT 0801, Australia.

Abstract

Dairy Arm is a small sand-bed stream whose channel and flood plain were essentially cleared of woody vegetation by early European settlers in the 1820s to 1880s for agriculture. Channel incision of Dairy Arm was initiated on 17-18 June 1949 by a catastrophic flood and resulted in substantial erosion of the channel in the upstream part of the channel network and massive deposition in downstream reaches. River training works carried out in the depositional zone between 1982 and 1991 predominately involved the installation of cylindrical wire mesh barriers for bank protection, stock-proof fencing and the planting of exotic poplars and willows and native river she-oaks. Thirteen permanently marked valley-floor cross sections were surveyed on three occasions between 1981 and 2009.

Quantitative data from the cross sections show that the incision cycle initiated by the 1949 flood essentially finished in the early 1990s. Recovery from incision is now well advanced and, in the former incised zone, involves channel contraction, bank revegetation (predominately grasses) and spatially discontinuous bed erosion. Bed erosion involves the creation of a pool-riffle sequence or chain of ponds, and is also very active in the former depositional zone. Channel width has contracted by the construction of a new bench and flood plain in the bed of the former incised channel. In 2009/2010, the vegetation on the bar/bed, bench and flood plain at six cross sections was sampled and the abundance of plant functional types examined. Floristic recovery on Dairy Arm is evidenced by the diversity of plant functional types present on fluvial landforms, by native and exotic species being evenly represented, and by the presence of species indicative of the reference plant community. A lack of catastrophic floods, the exclusion of livestock by river training works, presence of remnant vegetation in the upstream reaches of Dairy Arm (as well as where the channel abuts steep valley slopes), scattered shade trees retained on the flood plain, burial of introduced pasture by substantial sand deposition and the resulting reduction in stocking rates have all played a role in this recovery.

Whilst the abundance and diversity of herbaceous plants is good and they have played a crucial role in channel contraction, the same is not the case for woody species. Woody plants are present on the benches, banks and flood plain, but their abundance appears to be limited by grazing and the lack of supply of propagules. The absence of overstorey vegetation is hindering the recovery of important ecological functions, such as the regulation of stream temperature and the supply of large wood for further pool development. Full recovery of both the geomorphic processes and floristic diversity on Dairy Arm will take a long time yet because of the substantial lag time in natural recruitment from planted or remnant vegetation and the production of large wood from overstorey vegetation. This study shows that partial geomorphic and floristic recovery of an incised channel is possible in an agricultural setting, where grazing has been intermittent and stocking densities are low. However, progression to full recovery would be greatly enhanced by protecting the woody vegetation regenerating on the benches, banks and flood plain from livestock.

Keywords: channel recovery, channel contraction, pool reformation, plant functional types, grasses, woody vegetation, river she-oaks.

INTRODUCTION

The causes of channel incision and arroyo cutting have generated much controversy where grazing, cultivation, road construction and flood plain drainage replaced indigenous land uses in the eighteenth, nineteenth and/or twentieth centuries in various continents outside of Europe (for example, Cooke and Reeves 1976; Hendrickson and Minckley 1984; Schumm et al. 1984; 1987; Erskine 2005). Furthermore, recovery from channel incision has been attempted in many areas by simply reversing the assumed cause of incision (Cooke and Reeves 1976) although success rates have often been low (Peterson 1950; Peterson and Hadley 1955). Given the complexity of causal mechanisms of channel incision (Cooke and Reeves 1976; Hendrickson and Minckley 1984; Schumm et al. 1984; 1987; Erskine 2005), the identification of the correct cause of channel incision is difficult (Cooke and Reeves 1976). In this chapter, we report the rarely recorded partial geomorphic and floristic recovery of a former incised channel, Dairy Arm (Figure 1), over the last twenty years.

Figure 1. The Dairy Arm drainage basin in the upper Wollombi Brook catchment, Australia, showing locations of the cross sections.

Erskine and Melville (2008) demonstrated that the catastrophic flood of 17-18 June 1949 initiated historic channel incision of the upper drainage network of

Wollombi Brook in the Hunter Valley, NSW. Erskine (1996) found that Wollombi Brook has high flood variability by world standards, with Flash Flood Magnitude Indices (after Baker 1977) greater than 0.7. The flood of 17–18 June 1949 was the largest since first European settlement in the 1820s (Erskine 1994; 1996; Erskine and Melville 2008; Erskine and Chalmers 2009) but was greatly exceeded by at least three palaeofloods during the mid- to late-Holocene (Erskine and Peacock 2002). The highest rainfall for 17–18 June 1949 was 508 mm (Erskine 1994; 1996) at the head of the Dairy Arm basin. Flood peak discharges ranged between 22 and 39 times greater than the mean annual flood, depending on the gauging station and period of record used for flood frequency analysis (Erskine 1994; 1996; Erskine and Peacock 2002; Erskine and Chalmers 2009). The biogeomorphic processes of channel recovery from incision and their controlling factors are largely unknown. The purpose of this chapter is to address these biogeomorphic processes and understand how they are related to channel recovery.

Study Area

The study area consisted of pastoral land in the riparian corridor of Dairy Arm, which is an upper tributary of Wollombi Brook in the Hunter Valley, NSW, Australia (151°10'E, 33°02'S) (Figure 1). Dairy Arm has a total catchment area of 39.8 km^2. There are many discontinuously incised channels in the Dairy Arm basin (Figure 1). The alluvial section of Dairy Arm has been split into the following three zones based on their behaviour since European settlement, which are similar to those on the neighbouring Wollombi Brook (Erskine 2008). The *upper intact zone* extends 300 m upstream of cross section 1 (Figure 1) and covers the alluvial channel where it is unaffected by post-1949 incision. Valley width is constricted by the resistant Mangrove Sandstone Member of the Triassic Terrigal Formation (Hanlon et al. 1954; Uren 1974). A small capacity, sand-bed channel is flanked by a low floodplain covered by a moist forest with *Eucalyptus amplifolia* emergents. There is also an intact zone on Cullys, Smiths, McDonalds, Mahers and McMullens Arms but Olney Arm has incised along the whole alluvial reach (Figure 1).

The *incised zone* extends 5.7 km from cross section 2 to 9 on Dairy Arm (Figure 1) and has greatly widened and incised since 1949 (Erskine and Melville 2008). A series of knickpoints or headcuts are present between cross sections 1 and 3 along with log steps which produce organic stepping (Figure 2). Log steps are small dams formed of large wood (logs, limbs and/or roots > 0.1 m in diameter) which extend across the entire active channel (Figure 2), resulting in an abrupt drop in water surface elevation by overflow of the log (Marston 1982). They mark a section of channel at the upstream limit of historical channel incision (Figure 2). Knickpoints also mark the upstream limit of the incised zone on Cullys, Smiths, McDonalds, Mahers and McMullens Arms (Figure 3). All these tributaries join Dairy Arm as incised streams. Riparian vegetation has been extensively cleared downstream of cross section 3 on Dairy Arm and now consists of acacia woodland (*A. parramattensis* and *A. parvipinnula*). The incised channel is very wide and deep, and is flanked by extensive sandy benches and bars.

Figure 2. (A) Log step in bed of Dairy Arm near cross section 2, (B) Knickpoint in bed of Dairy Arm between cross sections 2 and 3.

Figure 3. Upstream extension of channel incision by knickpoint migration on McMullens Arm by one flood in October 1985 which was generated by a rainfall of 75.8 mm in one day. Left photo is before and right photo is after the October 1985 flood.

Figure 4. (A) Dairy Arm at the junction with Olney Arm in April 1978, (B) Dairy Arm at and upstream of cross section 12 in April 1978. Sand splays were deposited by the March 1978 flood. Cross section locations are shown in Figure 1. Photos obtained from the late John Gardner of the former NSW Water Resources Commission.

The *depositional zone* extends 5 km from cross section 9 to the junction with Wollombi Brook (Figure 1) and has experienced substantial sedimentation since 1949 (Figure 4) because of massive upstream channel erosion (Erskine and Melville, 2008). We started working on Dairy Arm immediately after the large flood of March 1978 which extensively eroded the incised zone and deposited massive sand splays across the whole floodplain in the depositional zone (Figure 4). The channel is much wider and shallower in the depositional zone than in the incised zone.

In May 1982, the then NSW Water Resources Commission commenced limited structural river training works and more extensive stock-proof fencing of willow, poplar and river she-oak plantings along the lower 5.5 km of Dairy Arm downstream of the Cullys Arm junction on the channel alignment and width created by the March 1978 flood (Water Resources Commission 1981). These works continued until 1991 and have largely been successful. The areas subject to stock-proof fencing have experienced supplemental plantings of native tree and shrub species by local landowners. Current stocking rates are low at 0.6 cattle per hectare. Similar stocking rates have been maintained since 1979.

METHODS

We have observed and measured channel changes on Dairy Arm and its tributaries since 1979. Quantitative data on recovery from channel incision have been obtained from all of the following sources. Thirteen permanently marked valley-floor cross sections were surveyed on Dairy Arm between 1981 and 1983. Various cross sections were resurveyed in 1985, 1996-7 and 2008-9. Five cross sections were surveyed on Cullys Arm in 1985 and 2009 (Figure 1). Longitudinal bed profiles centred on the cross sections were also surveyed at the same time. Bulk bed-material samples were collected at each section at the time of survey. Samples were air-dried and sieved at $\varphi/2$ intervals. Trenches were dug through the floodplain at selected cross sections and the sediments and sedimentary structures described.

Plant species and their abundance were determined within 5 m × 2 m quadrats, oriented with the longest axis parallel to the stream, at six cross sections (5, 6, 7, 10A, 10B and 13A). Cross sections with an A suffix are located where shown in Figure 1. Cross sections with a B suffix are located about 100 m downstream of the same numbered section with an A suffix (Figure 1). Sampling for vegetation composition occurred in November 2009 (except for cross section 13A which was sampled in January 2011). Quadrats were systematically placed in the bed, on both banks and within fenced river training works (where present) and, therefore, represented bed/bar, bank, bench and flood plain habitats. Two quadrats were placed in the bed at cross section 13A, as the bed was sufficiently wide to allow independence of the samples, whilst at cross section 5 one quadrat was placed in the bed and one on a point bar. There were only two bank quadrats available for analysis, so they were grouped with the bench quadrats. Thus, the replication for the bed/bar habitat was eight, whilst that for the benches and the flood plain was 12. In each quadrat, we recorded percent cover of all vascular plant species in the following cover classes: < 5%, 5-25%, 26-50%, 51-75%, 76-95% and 96-100%. The presence of "additional" woody species within a 10 m wide transect on both sides of each cross section (100 m in length) were systematically recorded, as well as those observed opportunistically along Dairy Arm.

Cross sections 5, 6 and 7 were unfenced and lightly grazed. At cross sections 10A and 10B, livestock were excluded from the bed and benches, but the flood plain was open to grazing. At cross section 13A stock could access all landforms through the unmaintained river training fences, but stocking densities were low. The most recent flood before vegetation sampling was in June 2007. Cover and type of bare ground (i.e. bed, bench or recent sand splays), leaf litter, large wood and projective foliage cover (i.e. canopy cover) were recorded. For all variables, averages were determined by substituting the midpoint of each cover class (Elzinger et al. 2001). Each species was allocated to *a priori* defined plant functional type (*sensu* Lavorel et al. 1997) on the basis of life-form (i.e. ferns, sedges, rushes, clonal grasses, tufted grasses, forbs, shrubs, trees). The woody, scrambling *Rubus* species were categorised as shrubs. The relative cover of each functional type was determined by summing the cover of species belonging to a particular functional type and expressing as a percentage of the total cover within the quadrat. Species nomenclature, life-form and country of origin follow Wheeler et al. (2002) for grasses and Harden (1990; 1991; 1992; 1994) for other species.

In August 2011 the adult stands of river she-oaks (*Casuarina cunninghamiana*) planted as part of the river training works were surveyed, as well as the seedlings within the channel. The five adult stands all occur downstream of cross section 10B and will hereafter be referred to as Stands 1 to 5. The stands consist of a single line of trees, ranging in length from 35 – 148 m. The diameter at breast height (1.3 m above the ground) of each individual was recorded, as well as whether it was in fruit. At each stand, three quadrats (10 m x 10 m) were placed in the channel with one directly adjacent to the stand, one upstream (20 m from the upstream edge of the canopy) and one downstream (20 m from the downstream edge of the canopy). Seedlings within the quadrats were counted and scored as either grazed or not grazed.

RESULTS

Recent Channel Changes

The purpose of this section is to quantify and explain recent channel changes.

Upper Intact Zone

This zone is now very short (<300 m) being sandwiched between the downstream incised zone and an upstream bedrock channel where Dairy Arm crosses the resistant Mangrove Sandstone Member of the Triassic Terrigal Formation (Hanlon et al. 1954; Uren 1974).

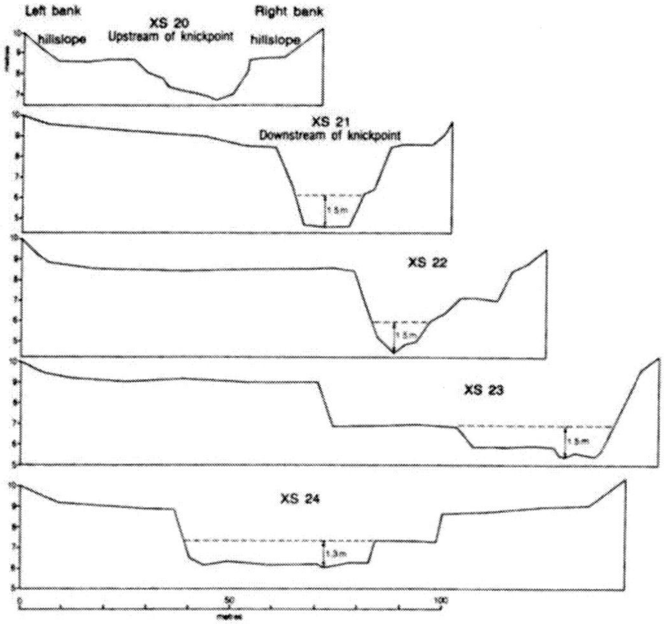

Figure 5. Downstream changes in valley-floor morphology of Cullys Arm. Cross section 20 is located upstream of the primary knickpoint in the intact zone and cross sections 21 to 24 are located in the incised zone which is integrated with Dairy Arm. See Figure 1 for location of cross sections.

The length of this reach decreased incrementally as knickpoints eroded the bed as they moved upstream between 1979 and 1990. However, by 1996 all knickpoints on Dairy Arm had been replaced by stable log steps (Figure 2) and the length of this zone has been constant since then. Only a 300 m long section of remnant alluvial channel has failed to incise since 1949. On Olney Arm, post-1949 incision extended right through the alluvial valley until it reached a stable ingrown meander cave (Jennings 1970) cut into massive sandstone of the Mangrove Sandstone Member of the Triassic Terrigal Formation (Hanlon et al. 1954; Uren 1974). Such incision also rejuvenated many tributaries (Figure 1). On Cullys Arm, post-1949 incision only extended about 2 km upstream of Dairy Arm (Figure 1) and left a swampy valley floor above the incised zone (Figure 5).

Incised Zone

The catastrophic June 1949 flood started incision near cross section 9. Many floods in 1950 and two floods in eight days in August 1952 continued the incision (Erskine and Melville 2008). By 1956, the incised channel was 5 km long and every major and most minor tributaries had also incised in response to drainage rejuvenation (Erskine and Melville 2008) (Figure 1). The catastrophic flood of March 1978 was the erosional finale of channel incision, greatly widening the incised channel and causing further upstream extension of incision. Between 1979, when we started our research, and 1990, the incised zone on all channels continued to progress upstream by knickpoint migration, as illustrated in Figure 3 for McMullens Arm. Some knickpoints were as high as 1.5 m and were usually capped by a resistant mud unit which overlay cross-bedded medium sand. As the main channel deepened, tributaries continued to be rejuvenated. Knickpoints are currently present on most tributaries in the incised zone (Figure 1). However, by 1996, this incision had moved only an additional 40 m upstream from where it was in 1985 on Dairy Arm, despite a large storm in February 1990 which caused erosion on many neighbouring alluvial fans (Scott and Erskine, 1994). By 1996, resistant logs had accumulated in such large numbers at the upstream end of this reach that stable log steps were very common. Log steps have contributed to the cessation of incision because knickpoints have now become ensconced on recently formed log steps. The incised channel margins and floodplain are now well vegetated, as discussed below. Bed degradation of up to 0.6 m has been active (Figure 6) since our first surveys, new floodplains have been deposited since 1982 in the incised trough eroded after 1949 (Figure 6) and bed-material grain size has significantly increased since 1996/7.

Major incision has now stopped in the incised zone. The March 1978 flood had a maximum daily rainfall in the Dairy Arm catchment of 166.2 mm and produced major upstream migration of existing knickpoints and channel widening. However, a larger event in June 2007 had a maximum daily rainfall of 185 mm at the same rainfall station (Yallambie) and produced extremely minor erosional impacts in the incised zone. The June 2007 flood was probably the largest flood on Dairy Arm since 1952 but had relatively minor geomorphic impacts on the channel network, further indicating that recovery from channel incision had commenced. The March 1978 flood was the erosional finale on Dairy Arm.

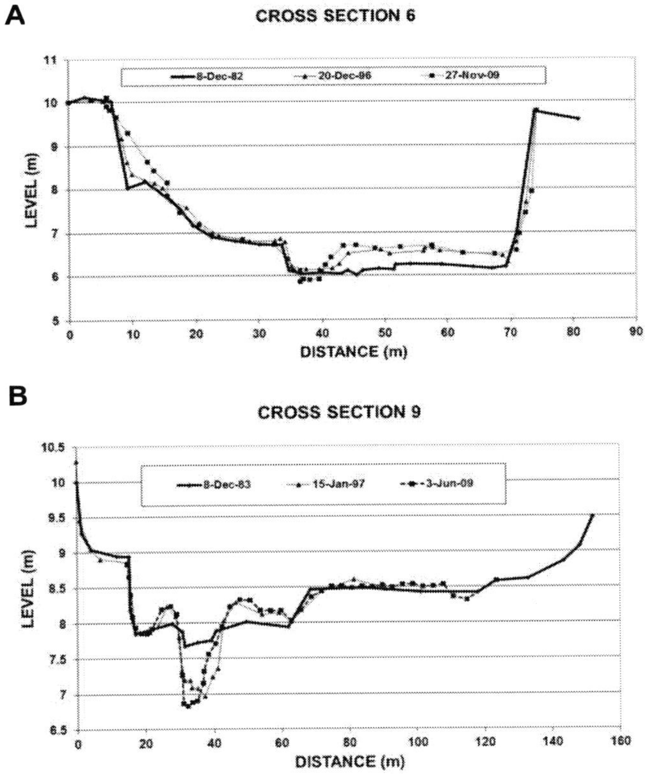

Figure 6. (A) Flood plain formation within the bed of the incised channel since 1982 at cross section 6 on Dairy Arm, (B) Bed degradation and flood plain formation within the bed of the incised channel since 1983 at cross section 9 on Dairy Arm. Location of cross sections shown in Figure 1.

Depositional Zone

The catastrophic March 1978 flood also greatly aggraded the bed and deposited extensive sand splays over the entire floodplain in the depositional zone (Figure 4). Channel width is currently less than half of what it was in March 1978, the bed has incised by up to 1.5 m and an extensive bench now flanks the channel inside the alignment fenced by the then NSW Water Resources Commission in the 1980s (Erskine et al. 2010).

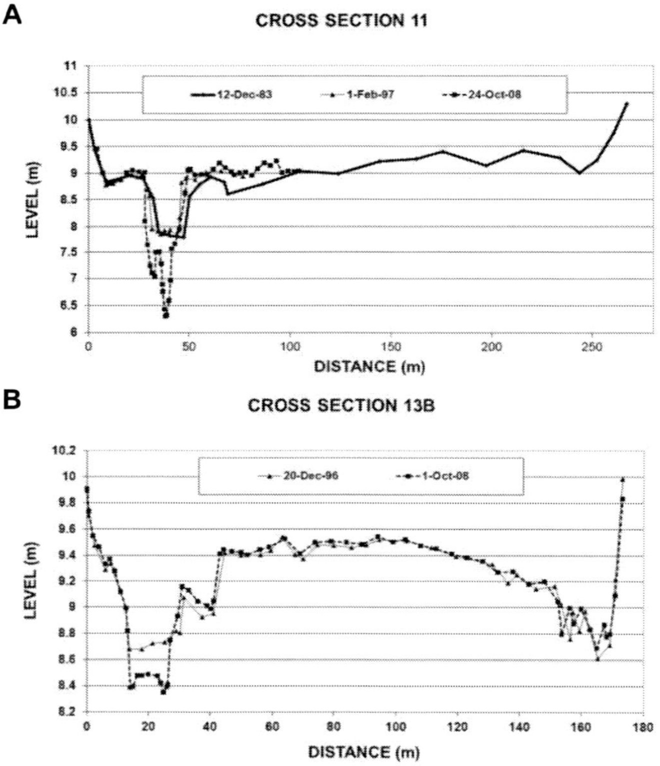

Figure 7. (A) Bed degradation and flood plain formation in the depositional zone since 1983 at cross section 11 on Dairy Arm, (B) Bed degradation in the depositional zone since 1996 at cross section 13B on Dairy Arm. Location of cross sections shown in Figure 1.

Figure 7 shows recent channel changes of Dairy Arm at cross sections 11 and 13B. While massive sand storage was apparent after the June 1949 flood and up to the March 1978 flood (Erskine and Melville, 2008), channel contraction and bed degradation are now active (Figure 7). Figure 8 demonstrates that the channel at cross section 13B has contracted since the construction of the river training works in 1982. Bed degradation between 1983 and 2008 has also caused an increase in mean bed-material size. Pools have now formed between cross sections 9 and 11 in an area where chain of ponds (Eyles 1977a; 1977b) were present before 1949. Some pools are very irregular and eroded into the fine-grained valley-fill that formed the valley-floor surface before the 1949 flood (Erskine and Melville 2008). These irregular pools are very similar to the chain of ponds described by Eyles (1977a; 1977b).

Figure 8. Upstream view to cross section 13B. (A) January 1981 before construction of river training works, (B) August 1982 after construction of river training works, (C) December 1985, (D) December 1996, (E) 18 April 2006, (F) 9 August 2008 after June 2007 flood. Compare (E) and (F) to determine minor biogeomorphic effects of June 2007 flood.

The road crossing between cross sections 11 and 12 (Figure 1) was completely buried with sand in 1978 but degradation since 1981 has now progressively exposed more of the formerly buried concrete causeway. The vertical drop over the causeway to the downstream bed is now 1.5 m (Figure 9). Downstream progressing bed degradation and subsequent pool formation will continue further downstream as upstream sand supply continues to decline (Erskine et al. 2010). Some pools now contain permanent water which was not the case before 2000.

Figure 9. Concrete road causeway located between cross sections 11 and 12 which was completely buried with sand until the late 1980s. It has been exposed continuously since then. Photo taken in February 1997.

Riparian Vegetation Composition and Regeneration

The purpose of this section is to quantify and describe the current composition of the vegetation and extent of regeneration. The canopy cover provided by shrubs and trees along the surveyed cross sections was low, being < 10% on all landforms (Table 1). As expected, the percentage of bare ground decreased and herbaceous ground cover and leaf litter increased with increasing distance from the channel bed (Table 1). The herbaceous cover consisted of a mixture of ferns, forbs, grasses, rushes and sedges. The percentage cover of woody debris was low (Table 1) and was predominately supplied by dead acacias.

A total of 38 plant species was recorded within the quadrats in the bed/bar habitat, whilst 54 species were recorded in the benches and the flood plain quadrats (Table 2). Overall, native and exotic species were evenly represented in the bench and flood plain habitats, whilst the number of natives exceeded the exotics in the bed/bar habitat. Forbs contributed the most to species richness on all landforms, followed by the grasses on the benches and flood plain and rushes/sedges on the bed/bar habitats (Table 2). However for the forbs, exotics out-numbered natives, especially on the benches. The number of native grass species either exceeded (bed/bar and bench habitat) or equalled (flood plain habitat) the number of exotic grass species (Table 2).

Table 1. Percentage cover of canopy cover of woody species, bare ground, herbaceous ground cover, leaf litter and woody debris on the three fluvial landforms examined on Dairy Arm. Number of samples (n) for each landform are given in parentheses

	Bed/bar (n = 8)	Bench (n = 12)	Flood plain (n = 12)
Canopy cover	0	3 ± 3.2	7 ± 5.3
Bare ground	55± 14.6	13 ± 3.9	8 ± 5.1
Ground cover	23 ± 11.5	63 ± 9.2	74 ± 7.5
Leaf litter	2 ± 0.5	9 ± 2.1	16 ± 3.5
Woody debris	2 ± 1.9	0.4 ± 0.28	2 ± 1.3

Table 2. Number of native and exotic species represented by each life form on the three fluvial landforms examined on Dairy Arm. Number of samples (n) for each landform are given in parentheses

Life form	Bed/Bar (n = 8)			Bench (n = 12)			Flood plain (n = 12)		
	Native	Exotic	All	Native	Exotic	All	Native	Exotic	All
Tree	0	0	0	1	1	2	1	2	3
Shrub	1	0	1	3	0	3	4	2	6
Fern	1	0	1	3	0	3	3	0	3
Forb	8	9	17	7	13	20	10	15	25
Grass	8	1	9	10	8	18	6	6	12
Rush/ sedge	5	5	10	5	3	8	4	1	5
Total	22	16	38	28	26	54	28	26	54

Exotic and native rushes/sedges were evenly represented in the bed/bar habitat, while the proportion of rush/sedge species that were exotic was lowest on the flood plain (Table 2). Ferns were present on all three landforms and all five species recorded were native (Table 2).

Woody species were poorly represented along Dairy Arm (Table 2; Figure 10). The only native tree species recorded in the quadrats or transects were juvenile eucalypts, which had very low cover (< 0.5%). The other two tree species were the exotic *Salix* sp. (1% cover on the benches and 6% cover on the flood plain) and *Populus nigra,* which was also present (0.2 % cover) on the flood plain. The shrub species were predominately native (Table 2), except for *Rubus anglocandicans* and *Solanum mauritianum* which were recorded in low abundance (0.2% cover) on the flood plain. The dominant native shrub species were *Acacia parvipinnula* (4% cover) and *Melaleuca decora* (3% cover), which were most abundant on the benches. *Rubus parvifolius*, a native, was the most abundant (5% cover) shrub species on the flood plain.

The relative cover of forbs was greatest on the bed/bar habitat, and decreased with increasing distance from the channel (Figure 10). Ferns were present on the bed/bar and bench/bank habitats, but were most abundant on the floodplain. Clonal grasses were abundant on all three fluvial landforms, but made up the greatest proportion of the vegetation on the bench/bank habitats (Figure 10). In contrast, the relative cover of tufted grasses declined with decreasing distance from the channel (Figure 10). Rush species were present on all three fluvial landforms but, not surprisingly, their relative cover was greatest in the bed/bar habitat (Figure 10). Sedge cover showed a similar trend, but they made up a far smaller proportion of the vegetation than rushes in the bed/bar habitat (Figure 10).

The composition of the remnant vegetation is consistent with it being Wollombi Alluvial Red Gum - Apple Forest as defined by Peake (2006). Therefore, the species listed as commonly occurring in Wollombi Alluvial Red Gum - Apple Forest are indicative of the reference vegetation community. Of the 26 native plant species listed by Peake (2006), 14 (54%) were observed along the riparian corridor of Dairy Arm during the current study (Table 3). To our knowledge only one of these species, *Casuarina cunninghamiana* (river she-oak), was planted as part of the river training works or subsequently by landholders.

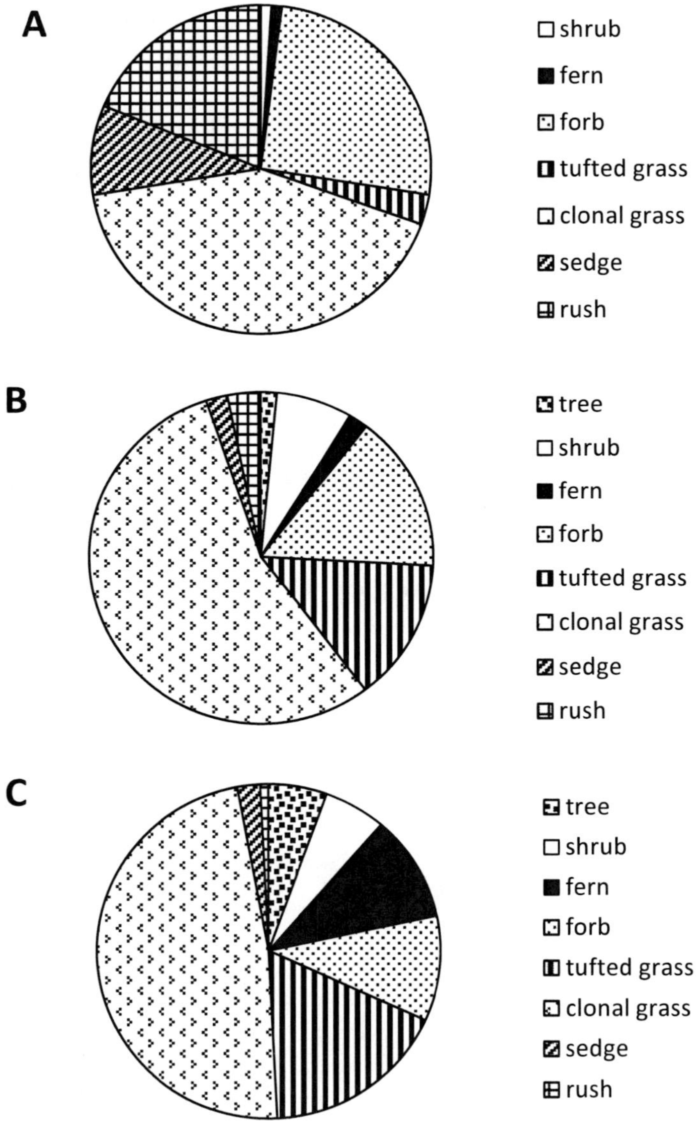

Figure 10. Proportional cover of each plant functional type on the (A) bed/bar, (B) bench/bank and (C) flood plain habitats on Dairy Arm.

Table 3. Species listed by Peake (2006) as commonly occurring in Wollombi Alluvial Red Gum – Apple Forest and their presence/absence along Dairy Arm (inside and outside quadrats) during this study

Family	Species	Presence/absence
Dennstaedtiaceae	*Pteridium esculentum*	+
Poaceae	*Echinopogon ovatus*	+
	Entolasia marginata	+
	Microlaena stipoides	+
	Oplismenus aemulus	-
Asteraceae	*Sigesbeckia orientalis*	-
Casuarinaceae	*Casuarina cunninghamiana*	+
Commelinaceae	*Commelina cyanea*	+
Convolvulaceae	*Dichondra repens*	+
Fabaceae	*Acacia parramattensis*	+
	Acacia parvipinnula	+
Lamiaceae	*Clerodendrum tomentosum*	-
Meliaceae	*Melia azedarach*	-
Menispermaceae	*Stephania japonica*	-
Myrtaceae	*Angophora floribunda*	+
	Backhousia myrtifolia	-
	Eucalyptus amplifolia	+
	Eucalyptus eugenioides	-
	Eucalyptus tereticornis	-
	Leptospermum polygalifolium	+
Phyllanthaceae	*Phyllanthus gunnii*	-
Pittosporaceae	*Bursaria spinosa*	+
Rosaceae	*Rubus parvifolius*	+
Ranunculaceae	*Clematis glycinoides*	-
Sterculiaceae	*Brachychiton populneus*	-
Ulmaceae	*Trema tomentosa*	-

The average diameter at breast height (DBH) of the planted river she-oak stands ranged from 21 to 39 cm (Table 4). About half of the planted trees were in fruit during the study, except for stand 5 where only one of the five individuals was reproductive. The number of seedlings in the bed/bar habitats was spatially variable (patchy) and appeared unrelated to the number of adults in fruit (Table 4). The proportion of seedlings that showed evidence of being grazed ranged from 11 – 100% (Table 4). However this upper figure is not neccessarily indicative of the grazing pressure at stand 1, as only five seedlings in total were recorded in the quadrats. The majority of river she-oak individuals in the channel were < 0.3 m in height (data not shown) and only a few juveniles greater than 1 m tall were observed on any landform along Dairy Arm.

Table 4. Characteristics of the adult stands of *Casuarina cunninghamiana* (river she-oaks) planted as part of the river training works and seedling regeneration within the channel (i.e. bed and bars). DBH = diameter at breast height

	Stand 1	Stand 2	Stand 3	Stand 4	Stand 5
Number of adults fruiting/total	12/17	5/10	9/15	13/23	1/5
Mean (± SE) DBH of adults (cm)	36 ± 3.0	37 ± 3.9	36 ± 4.3	21 ± 2.9	39 ± 5.6
Mean (± SE) number of seedlings/100 m^2	2 ± 1.7	33 ± 16.9	6 ± 5.7	126 ± 113.5	45 ± 20.2
% seedlings grazed	100	11	53	26	13

DISCUSSION

A substantial reduction in sand supply to the channel has occurred in the Dairy Arm catchment since the catastrophic flood of March 1978 (Erskine et al. 2010). This reduction has been achieved by a combination of channel incision progressing as far upstream as it is likely to reach under existing bioclimatic conditions and by vegetation invasion of the river banks, marginal benches and floodplain by stoloniferous and rhizomatous grasses. Some structural works, stock-proof fencing and plantings assisted vegetation recovery. Erskine et al. (2010) found a non-significant decrease in mean annual rainfall of between 9 and 21% at four rainfall stations since 1990, which has reduced the frequency of flood disturbance and enhanced the survival of grasses and scattered acacia individuals. This vegetation has now stabilised the channel banks, effectively stopping most bank erosion. Reduced sand supply, particularly since 1990, has induced a secondary phase of bed degradation which is much less severe than the primary incision initiated by the catastrophic June 1949 flood (Erskine and Melville, 2008). Furthermore, the development of a new pool-riffle sequence in the depositional zone has commenced and is providing aquatic habitat and water storage for the first time since the incision-initiating flood of June 1949. Sediment reduction is essential for the recovery of incised channels because episodic pulses of large volumes of sand are generated by catastrophic floods and overwhelm the channel in areas subject to high flood variability (Erskine 1994; 1996; 2008). While sediment supply will eventually decline naturally once all alluvial reaches are incised, vegetation in conjunction with structural works can be used to stabilise banks and knickpoints, and hence assist channel recovery during periods of little flood disturbance. Methods of predicting periods of little flood disturbance are essential for river restoration in Australia.

The remnant vegetation observed within the riparian corridor of Dairy Arm during the current study suggests that Myrtaceous species such as *Angophora floribunda* (rough-barked apple), *Eucalyptus amplifolia* (cabbage gum), *Leptospermum polygalifolium* (lemon-scented tea-tree) and *Melaleuca decora* (white feather honeymyrtle) have been largely cleared from the flood plain and banks of Dairy Arm and the forest replaced by an acacia woodland (Figure 11A). The high cover of herbaceous species from the Cyperaceae, Juncaceae, and Poaceae families found on the bed/bar and bench habitats is most likely due to the high light levels associated with the absence of a forest canopy. This is consistent with the changes in the pollen record found by Bennett and Mooney (2003) in cores taken from backswamps adjacent to Wollombi Brook of which Dairy Arm is a tributary.

Bennett and Mooney (2003) showed that pollen counts of Myrtaceous and *Casuarina* species declined, while counts of *Acacia*, Asteraceae, Cyperaceae and Poaceae species increased after European settlement. They also found that fern spores declined, with the exception of *Pteridium* (bracken) which increased.

For a riparian corridor with a long history of grazing, the number and abundance of native grasses, sedges and rushes in the current study were higher than expected.

Although the banks and flood plain along Dairy Arm have been previously sown to paspalum (*Paspalum dilatatum*) and white clover (*Trifolium repens*), two large floods deposited sand with a combined depth of 1 m, completely burying the pasture (Erskine and Melville 2008). As the paspalum and white clover could not recover, the pasture supported few cattle and the stocking rates were reduced (N. Thompson pers. comm. 2011).

Stocking rates remain low, as the quantity and quality of grasses are unable to maintain stock condition in winter and the cattle require supplementary feed (N. Thompson pers. comm. 2011).

Thus channel incision initiated upstream by the catastrophic 1949 flood, and which resulted in massive deposition on the lower reaches of Dairy Arm, appears to have inadvertently led to conditions likely to favour native herbaceous species, i.e., the burial of competitive, exotic pasture species with nutrient-poor sediment and a reduction in livestock numbers.

Figure 11. (A) woodland dominated by acacias near cross section 6, (B) regenerating cabbage gums on the right bank near cross section 12. Photos taken between 2009 and 2011.

One of the reasons paspalum may not have been able to survive burial by sand, or rapidly colonize from adjacent areas, is that it is not clonal. In contrast, the rhizomes and stolons of grasses such as *Cynodon dactylon* (couch), *Isachne globosa* (swamp millet), *Paspalum distichum* (water couch) and *Phragmites australis* (phragmites) enables their recovery after burial by sediment deposited by floods. Our results show that clonal grasses (Figure 12A) were much more abundant than tufted grasses in bed/bar habitats (where sand is highly mobile and constantly being re-worked) and also in bench/bank habitats (where sand splays are often deposited). Our results also show that the structural diversity of herbaceous plants within the bed and at the water's edge is high, with ferns as well as clonal grasses, tufted grasses, sedges, rushes and forbs being present. Within the bed of the former depositional zone, scouring near and around some rush/sedge species was noted and, in some cases, gravels were exposed (Figure 12B). This suggests that this group of species may function to increase in-channel habitat.

Figure 12. (A) clonal grasses (water couch and swamp millet) in the bed near cross section 6, (B) scour around sedge exposes gravel near cross section 12, (C) remnant vegetation dominated by cabbage gums (left) and acacias (right) upstream of cross section 4, (D) River she-oak seedlings on a low bench upstream of adult planting near cross section 13B. Photos taken between 2009 and 2011.

Whilst the diversity and cover of shrub species is low, they were at least predominately native. Shrub species with soil stored seed banks (i.e. *Acacia* sp.) were most common. Although the seed longevity of *A. parvipinnula* and *A. parramattensis* is unknown, research shows that seed of a range of Australian acacias lose little viability over (dry) storage periods of 18 years (Langkamp 1987), and, in exceptional cases, may still be viable after 151 years (Leino and Edqvist 2010). However moist storage conditions cause drastic loss in viability (Langkamp 1987), so these studies may overestimate seed longevity *in situ*. During the current study, seedlings of *Acacia* species were recorded on all fluvial landforms and anecdotal observations at the junction of Dairy Arm with Wollombi Brook suggests that juveniles on benches can withstand their stems being buried with sand and form adventitious roots on the stem. The acacias on Dairy Arm are known to recruit on masse after flood (e.g. after the March 1978 flood). It is also possible that the annual pastoral management burn stimulates the soil stored seedbank. Nitrogen fixing genera, such as *Acacia* and *Casuarina,* may play an important ecosystem function by improving nitrogen availability in sand deposited by large floods on the benches and banks of sand-bed streams and, in the long-term, this may facilitate the survival of later successional species.

The lack of eucalypt regeneration in former agricultural areas in Australia has been widely documented (e.g. Weinberg et al. 2010). Although woody species with canopy stored seedbanks (predominately dry-fruited Myrtaceous genera such as *Angophora, Eucalyptus, Leptospermum* and *Melaleuca*) are present along Dairy Arm recruitment appears to be limited by the supply of propagules and grazing. Although there is an upstream source of cabbage gum (Figure 12C) and scattered individuals surviving on the flood plain margins, there is little natural regeneration. This is consistent with studies showing that the dispersal distances of eucalypt seed is relatively short and usually less than double the height of the tree canopy (McDonald 1999). Some regeneration of cabbage gum has been observed in a few places where individual trees have reached maturity on the bank (Figure 11B). It is unknown whether these few mature cabbage gum individuals were planted by landholders on the bank of Dairy Arm or whether the seed dispersed naturally via hydrochory. Where the channel abuts the base of the steeper valley slopes, and remnant or regrowth forest overhangs the creek, seedlings of cabbage gum are present on a bench within the channel. The likelihood of these individuals surviving to adulthood will depend on the timing of the next large flood.

Scattered individuals of white feather honeymyrtle have been retained on the flood plain as shade trees for cattle (N. Thompson pers. comm. 2011). Regeneration of white feather honeymyrtle has been observed on the bars and benches within the channel over a limited area close to where a small number of individuals have reached maturity on high benches (Figure 13A). Lemon-scented tea-tree has also been observed regenerating on a low bench in one location. It is unknown whether these mature individuals of white feather honeymyrtle and lemon-scented tea-tree were planted by landholders to augment the earlier river training works or whether they have dispersed and regenerated naturally. These two shrub species are likely to enhance bench formation in a way similar to that reported by Erskine et al. (2009) for river she-oak.

Figure 13. (A) white feather honeymyrtle regenerating on the right bank bench near cross section 13B, (B) abandoned pond containing tall spike-rush and water ribbons. Photos taken between 2009 and 2011.

Out of the 300 river she-oak seedlings that were re-introduced to Dairy Arm in 1987 as part of the river training works, about 72 remain. The majority are reproductive, and prolific (but spatially patchy) recruitment of seedlings has been observed on the bed, bars and low benches within the channel (Figure 12D). Despite the availability of viable seed few juveniles have survived on higher benches and banks, whilst regeneration is absent from the flood plain. On another sand-bed stream in the Upper Hunter Valley, Erskine et al. (2009) concluded that stands of river she-oaks established on benches within the channel between December 1983 and September 1992 (around the time of river training plantings in this study) and have survived through to maturity. Considering that *Casuarina* species become reproductive at 4-5 years of age (Wheeler et al. 2011), seed from the trees planted at Dairy Arm would not have become available until about 1992. Erskine et al. (2010) reported a decline in mean annual rainfall between 1959-1990 and 1991–2008 within the Dairy Arm catchment. Thus, the stands that have reached reproductive maturity were planted on the banks during a wetter period (i.e. 1987), but seed production would have coincided with a drier period that may have been unsuitable for seedlings to establish.

Recovery Potential

Although the sources of the commonly occurring woody species typical of Wollombi Alluvial Red Gum – Apple Forest are present along the riparian corridor of Dairy Arm, those species with canopy held seed are often at distances from the channel that exceed wind dispersal distances reported for dry-fruited Myrtaceous species. Livestock grazing, seed predation by ants, and unsuitable climatic conditions may also be restricting regeneration, as has been demonstrated for river red gum (*Eucalyptus camaldulensis*) elsewhere in Australia (Meeson et al. 2002). While hydrochory is likely to be occurring, the rate of migration via this mechanism appears to be exceedingly slow. Although pool and riffle sequences have not fully recovered, abandoned pools adjacent to the current channel contain species typical of deeper water, such as *Eleocharis sphacelata* (tall spike-rush) and *Triglochin procera* (water ribbons), which will provide the necessary source populations if deeper pools develop in the channel in the future (Figure 13B).

Although about half of the species commonly found in Wollombi Alluvial Red Gum – Apple Forest currently occur on Dairy Arm, it is the species typically associated with humid and/or shaded environments (e.g. *Backhousia myrtifolia, Brachychiton populneus, Clematis glycinoides, Clerodendrum tomentosum, Oplismenus aemulus, Sigesbeckia orientalis* and *Stephania japonica*) that are missing. These species will only recover once the shading and microclimate provided by overstorey vegetation along the channel is re-established. Before attempting to reintroduce these "missing species", suitable reference sites need to be found to confirm the appropriateness of these species for Dairy Arm, the likelihood of natural dispersal (i.e. whether reintroduction is actually necessary) and the appropriate timing of any reintroductions.

Large wood loadings significantly influence the type and extent of pools present in Australian sand-bed forest streams (Webb and Erskine 2005; Erskine et al. 2012), but Vesk et al. (2008) estimated that the lag time in producing fallen timber loads similar to that of mature forest, from revegetation sites in central Victoria, takes about 150 years. We believe that in the current riparian zones of Dairy Arm large wood supply from revegetated channels and flood plains will take at least 50 years. Furthermore, one study has predicted that vegetation is needed along at least 2 km of stream to measurably alter water temperature (Rutherfurd et al. 2004). Thus, an absence of overstorey vegetation is hindering the recovery of important ecosystem functions along Dairy Arm, such as the regulation of stream temperature and micro-climate by shading, the supply of detritus for decomposers and aquatic macro-invertebrates, and the production of large wood for further pool development.

Although this study shows that partial geomorphic and floristic recovery is possible in an agricultural setting, where grazing has been intermittent and stocking densities are low, it will take a long time yet for full channel and floristic recovery. Progression to full recovery would be greatly enhanced by protecting the woody vegetation regenerating on the benches and banks from livestock.

Acknowledgements

We thank Mike Melville, Nev Thompson, Clare and Simon Hawse, Cyrus and Barry McMullen, Stan Brown and Dr Palmisano for their sharing of information and practical assistance. We would also like to thank Chelsea Zuiderwyk and Karen O'Neill for assistance with vegetation sampling, and Diane Warman for assistance with plant identification. Tyrone Robertson and Chris Myers assisted with surveys.

References

Baker V. (1977) Stream-channel response to floods, with examples from central Texas. *Geological Society of America Bulletin* 88, 1057–1071.

Bennett P. and Mooney S. D. (2003) A palynological reconstruction of pre-European riparian vegetation at Wollombi, New South Wales and its application to stream bank management and revegetation. *Ecological Management and Restoration* 4, S69-S78.

Cooke R. and Reeves R.W. (1976) *Arroyos and Environmental Change in the American South-West.* Clarendon Press, Oxford.

Elzinger C. L., Salzer D. W., Willoughby J. W. and Gibbs J. P. (2001) *Monitoring plant and animal populations.* Blackwell Science, Massachusetts.

Erskine W.D. (1994) Flood-driven channel changes on Wollombi Brook, NSW since European settlement. In: Mahony D. and Whitehead J. (Eds), *The Way of the River. Environmental Perspectives on the Wollombi.* Wollombi Valley Landcare Group and The University of Newcastle, Newcastle, 41-69.

Erskine W. D. (1996) Response and recovery of a sand-bed stream to a catastrophic flood. *Z. Geomorph. N.F.* 40, 359–383.

Erskine W.D. (2005) Gully erosion. In: J.H. Lehr, J. Keeley, J. Lehr and T.B. Kingery III (Eds), *Water Encyclopedia: Surface and Agricultural Water.* Wiley-Interscience, Hoboken, pp. 183-188.

Erskine W.D. (2008) Channel incision and sand compartmentalization in an Australian sandstone drainage basin subject to high flood variability. In: *Sediment Dynamics in Changing Environments* (ed. by J. Schmidt, T. Cochrane, C. Phillips, S. Elliott, T. Davies and L. Basher) (Proc. Christchurch Symp. Dec. 2008), 283-290. IAHS Publ. 325. IAHS Press, Wallingford, UK.

Erskine W.D. and Chalmers A. (2009) Natural river recovery from catastrophic channel changes by vegetation invasion of the sand-bedded Wollombi brook, Australia. *Proceedings of the 7th International Symposium on Ecohydraulics, 12-16th January 2009, Conceptión, Chile.*

Erskine W.D., Chalmers A., Keene A., Cheetham M. and Bush R. (2009) Role of a rheophyte in bench development on a sand-bed river in southeast Australia. *Earth Surface Processes and Landforms* 34, 941-953.

Erskine W.D., Chalmers A.C. and Townley-Jones M. (2010) The importance of sediment control for recovery of incised channels. In: *Sediment Dynamics for a Changing Future* (ed. by K. Banasik, A. J. Horowitz, P.N. Owens, M. Stone and D.E. Walling) (Proc. The Warsaw University of Life Sciences - SGGW, Poland, Symp. June 2010). IAHS Publ. 337, 2010.

Erskine, W.D. and Melville, M.D. (2008) Geomorphic and stratigraphic complexity: Holocene alluvial history of upper Wollombi Brook, Australia. *Geografiska Annaler* 90A, 19–35.

Erskine W. D. and Peacock C. T. (2002) Late Holocene flood plain development following a cataclysmic flood. In: *The Structure, Function and Management*

Implications of Fluvial Sedimentary Systems (ed. by F.J. Dyer, M.C. Thoms and J. Olley) (Proc. Alice Springs Symp. Sept. 2002), 177–184. IAHS Publ. 276. IAHS Press, Wallingford, UK.

Erskine W.D., Saynor M.J., Chalmers A.C. and Riley S.J. (2012) Water , wind, wood and trees: interactions, spatial variations, temporal dynamics and their potential role in river rehabilitation. *Geographical Research*, in press.

Eyles R. J. (1977a) Birchams creek: the transition from a chain of ponds to a gully. *Australian Geographical Studies* 15, 146-57.

Eyles R. J. (1977b) Changes in drainage networks since 1820, Southern tablelands, NSW. *Australian Geographer* 13, 377-86.

Hanlon F. N., Osborne G. D. and Raggatt H. G. (1954) Narrabeen Group: its subdivisions and correlations between the South Coast and Narrabeen-Wyong districts. *Journal and Proceedings of Royal Society of NSW* 87(3), 106-120.

Harden G.J. (1990) *Flora of New South Wales, Volume 1.* New South Wales University Press, Sydney.

Harden G.J. (1991) *Flora of New South Wales, Volume 2.* New South Wales University Press, Sydney.

Harden G.J. (1992) *Flora of New South Wales, Volume 3.* New South Wales University Press, Sydney.

Harden G.J. (1994) *Flora of New South Wales, Volume 4.* New South Wales University Press, Sydney.

Hendrickson D.A. and Minckley W.L. (1984) Cienegas – vanishing climax communities of the American southwest. *Desert Plants* 6, 131-175.

Jennings J. N. (1970) Australian Landform Example No. 18. Ingrown meander and meander cave. *Australian Geographer* 11, 401-2.

Langkamp P. (1987) *Germination of Australian Native Plant Seed.* Inkata Press, Melbourne.

Lavorel S., McIntyre S., Landsberg J. and Forbes T. D. A. (1997) Plant functional classifications from general groups to specific groups based on response to disturbance. *Trends in Ecology and Evolution* 12, 474 - 8.

Leino M. W. and Edqvist J. (2010) Germination of 151-year old Acacia spp. seeds. *Genetic Resources and Crop Evolution* 57, 741-6.

McDonald T. (1999) Strategies for the ecological restoration of woodland plant communities: harnessing natural resilience. In: *Temperate eucalypt woodlands in Australia: biology, conservation, management and restoration* (eds R. J. Hobbs and C. J. Yates) pp. 286-97. Surrey Beatty and Sons, Chipping Norton.

Marston R.A. (1982) The geomorphic significance of log steps in forest streams. *Annals of the Association of American Geographers* 72, 99-108.

Meeson N., Robertson A. I. and Jansen A. (2002) The effects of flooding and livestock on post-dispersal seed predation in River Red Gum habitats. *The Journal of Applied Ecology* 39, 247-58.

Peake T. C. (2006) *The vegetation of the Central Hunter Valley, New South Wales. A report on the findings of the Hunter remnant vegetation project.* Hunter-Central Rivers Catchment Management Authority, Paterson.

Peterson H. V. 1950. The problem of gullying in western valleys. In: Trask P. D. (Ed.), *Applied Sedimentation.* Wiley, New York, 407-434.

Peterson H. V. and Hadley R. F. (1955) Effectiveness of erosion abatement practices on semiarid rangelands in western United States. *International Association of Hydrological Sciences* 53, 182-191.

Rutherfurd I. D., Ladson A. R. and Stewardson M. J. (2004) Evaluating stream rehabilitation projects: reasons not to, and approaches if you have to. *Australian Journal of Water Resources* 8, 57-68.

Schumm S. A., Harvey M. D. and Watson C. C. (1984) *Incised Channels.* Water Resources Publ., Littleton, Colorado, USA.

Schumm S. A., Mosely M. P. and Weaver W.E. (1987) *Experimental Fluvial Geomorphology.* John Wiley and Sons, New York.

Scott P.F. and Erskine W. (1994) Geomorphic effects of a large flood on fluvial fans. *Earth Surface Processes and Landforms* 19, 95-108.

Uren R. E. (1974) The Geology and Low Cost Extractive Resources of the Gosford Lake Macquarie Area. Volume 1 Resources. NSW Department of Mineral Resources Unpublished Geological Survey Report 1974/431.

Vesk P. A., Nolan R., Thomson J. R., Dorrough J. W. and Nally R. M. (2008) Time lags in provision of habitat resources through revegetation. *Biological Conservation* 141, 174-86.

Water Resources Commission of NSW (1981) Proposal for the control of Dairy Arm in the vicinity of Yallambie. Submission to the Hunter Valley Conservation Trust, Maitland.

Webb A. A. and Erskine W. D. (2005) Natural variability in the distribution, loading and induced scour of large wood in sand-bed forest streams. *River Research and Applications* 21, 169-85.

Weinberg A., Gibbons P., Briggs S. V. and Bonser S. P. (2010) The extent and pattern of Eucalyptus regeneration in an agricultural landscape. *Biological Conservation* 144, 227-33.

Wheeler D. J. B., Jacobs S. W. L. and Whalley R. D. B. (2002) *Grasses of New South Wales.* University of New England, Armidale.

Wheeler G. S., Taylor G. S., Gaskin J. F. and Purcell M. F. (2011) Ecology and management of sheoak (*Casuarina* spp.), an invader of coastal Florida, USA. *Journal of Coastal Research* 27, 485-92.

In: Riparian Zones
Editor: A. Kerem et al.

ISBN 978-1-61942-677-1

Chapter 4

GROUND WATER CHEMISTRY UNDER RIPARIAN ZONES AND ITS POSSIBLE EFFECT ON THE LONGITUDINAL CHANGE IN STREAM WATER CHEMISTRY IN THE FORESTED WATERSHEDS, NORTHERN JAPAN

Yasunori Nakagawa*
7-32 Motomachi, Ebetsu City, Hokkaido Pref., 067-0031 Japan

ABSTRACT

I evaluated ground water chemistry under riparian zones and stream water chemistry in two forested watersheds, northern Japan. We selected two small watersheds; 'southern watershed' had an extending riparian wetland, and 'northern watershed' had a narrow riparian area. I measured the stream water chemistry from the upstream (spring) to the downstream (the outlet of the watershed). In longitudinal increase of the stream water flow, stream water chemistry also changed longitudinally in the watersheds. The concentrations of major ions except for ammonium ion (NH_4^+) in both watersheds and chloride ion (Cl^-) in the southern watershed decreased downstream. On the other hand, the concentration of Cl^- and dissolved iron (Fe) in the southern stream increased. NH_4^+ concentrations in the stream water in both watersheds were low (<35 μmolc/L) and did not significantly

* E-mail: y7045f-n@k2.dion.ne.jp

changed downstream. The downstream decreases of the most solutes were caused by the inflow of the surrounding ground water having low concentrations under the riparian zone to the stream channel. However, Fe concentration in the ground water was higher than in the stream water. No significant change of the Fe concentration in the northern stream suggested the Fe precipitation at the soil-stream interface. Moreover, dissolved organic carbon (DOC) would be related to the nitrogen transformation in the riparian soil and the releases of dissolved Fe, dissolved aluminum (Al) and dissolved silica (Si) from the riparian soil.

INTRODUCTION

Riparian zone is a contact between land and flowing water. Flowing water generates from the spring, and then increases water volume with involving the surrounding ground water; it becomes stream, and river further, finally reaches the ocean (Figure 1). Riparian zones are usually under wet conditions, and they are kinds of wetlands. However, the water existing under the riparian zones exchange the flowing water somewhat oxidized (Hedin et al. 1998). Thus, riparian zone is the different type of the wetland, and has the different regimes of the material cycling. The hyporheic zone, especially exhibiting significant surface-subsurface exchange, functions as an important control on nutrient dynamics in the riparian zone and streambed (e.g. Hill et al., 1998; Shibata et al., 2004; Triska et al. 1989).

We selected two small forested watersheds having different watershed area, elevation range, stream length, area of riparian zone, etc. (Table 1) to evaluate 1) the longitudinal pattern of the stream chemistry from the spring to the outlet of the watershed and 2) the effects of the riparian zone on this longitudinal pattern. Combining these two viewpoints, we evaluated the effects of the soil and ground water chemistry in the riparian zone on the stream water chemistry at a scale of a whole stream channel, and compared the two watersheds. Moreover, the survey of the spatial variety of the soil and ground water chemistry in the riparian zone may suggest which part of the riparian zone intensively affected the stream water chemistry.

In the past study, I focused on nitrate ion (NO_3^-), dissolved organic carbon (DOC), and dissolved iron (Fe) concentrations among many dissolved substances in the forested watersheds same as the present study (Nakagawa et al. 2008). I also focused on stream pH in the stream water (Nakagawa et al. 2009).

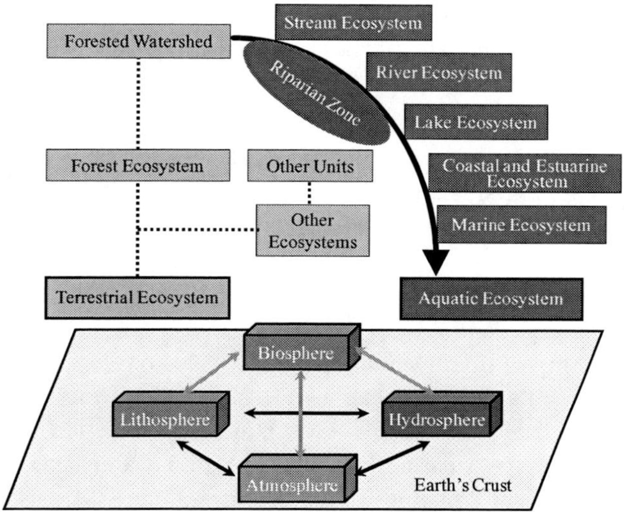

Figure 1. An environmental position of the Riparian Zone.

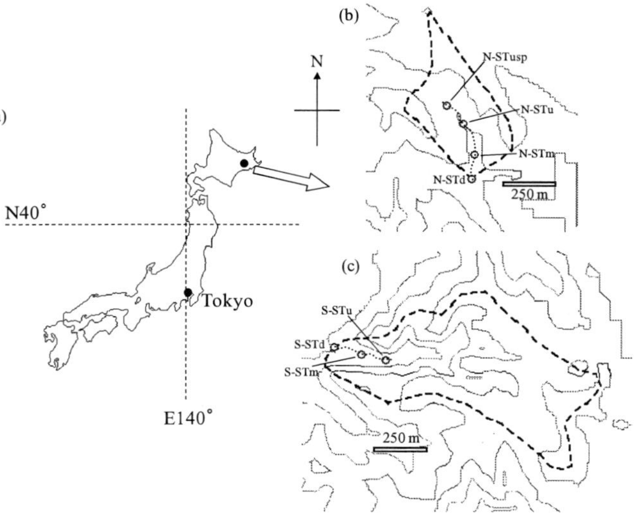

Figure 2. Study site. (A) location of the study site, (b) northern watershed, and (c) southern watershed. The open circles show the sampling point of the stream water. The characters before hyphens, "N" and "S", denote northern and southern watersheds, respectively. The characters after hyphens are as follows; STusp: upper spring, STu: upstream, STm: midstream, and STd: downstream. Contour lines are drawn at 20 m intervals. Dashed lines and dotted lines are watershed boundaries and stream channels, respectively.

The reasons for selecting these three substances were that 1) they were nutritionally (NO_3^- and Fe) and energetically (DOC) important for the downstream ecosystems, and that 2) their behaviors were closely related to the riparian reductive conditions.

NO_3^- is one of the major eutrophicating nutrients, and Fe is a limiting element for some maritime ecosystems (Martin and Fitzwater, 1988; Martin et al., 1990). DOC is an important substance for the heterotrophic activities and solar radiation inhibition in aquatic ecosystems. Moreover, DOC is generally abundant in the wetlands due to low decomposition rate (Dalva and Moore, 1991). Carbon accumulation on the wetland affects the mineral density in the soil layers, perhaps affecting the leaching of dissolved silica (Si). DOC also related to the leaching of aluminum (Al) and Fe (e.g. Kaneko et al. 2003). The leaching of these two elements was related also to the pH and Oxidation-Reduction Potential (ORP). Thus, the studies on the riparian zones in forested watersheds would be also important for the nutrient dynamics in the downstream ecosystems. In the present chapter, I reanalysis the data of the stream water chemistry, recompile the results of the above two studies (Nakagawa et al. 2008 for NO_3^-, DOC, and Fe; 2009 for pH), and then add the knowledge of the other elements.

Figure 3. Pictures of the watersheds studied.

METHODS

Site Description

We studied the chemistry of mountainous streams in two forested watersheds in the Hokkaido Forest Research Station, Field Science Education and Research Center (FSERC) Kyoto University, in Shibecha, eastern Hokkaido, northern Japan

(Figure 2). The office of this station is located at N 43°19′, E 144°37′, where meteorological data were obtained. The annual mean temperature and mean annual precipitation were 5.7 degrees centigrade and 1157 mm (1971-2000), respectively (FSERC, Kyoto University). The period of the present study was from August 2001 to December 2004. The annual mean temperature was 6.8 degrees centigrade and annual precipitation during the present study was 1045 mm (Table 1). Precipitation exhibited a relatively even distribution at the time of the present study, but was relatively high in August and September. In winter, snowfall was observed and usually about 25 % of annual precipitation. In the present site, seasonal soil freezing was observed from December to May. Maximum freezing depth varied from 35 to 50 cm and generally occurred in early March (Takeuchi 1980).

To simplify the effects of the riparian zone on the stream water chemistry, the selected watersheds had relatively short stream channels with less branching (second-order streams) and less meandering than the other watersheds in this region, although a few tributaries and some curves were unavoidable for selecting the stream. Table 2 shows the properties of the two watersheds. One watershed is named the ‘southern watershed’, and is defined as that having a broad and flat ‘riparian wetland’. This watershed is located 1.9 km northeastward of the above-mentioned office. Another watershed is named the ‘northern watershed’, and has a narrow ‘riparian area’. This watershed is located 8 km north-northeastward of the office.

The soil type was Humic Andosols (FAO/Unesco 1974) derived from volcanic ash of Masyu (Ma~f) in the southern watershed and of Masyu g-j (Ma-g~j) in the northern watershed; small characters a~f denote older material than g~j (Machida and Arai 1992). The depth of the Oi layer was 2-3 cm, and the sum of Oe and Oa layers was very thin (< 1 cm). The mineral soil layer was well-mixed with humus, and well-developed aggregates were recognized clearly in this layer. In the southern watershed, most of the wetland is overlain by a surface peat layer of 7 cm on the soil. Bedrock is Abukuma Granitoids including basalt and chert. The vegetation of both watersheds was natural deciduous broad-leaved forest.

The canopy layer was composed of *Ulmus davidiana* var. *japonica*, *Fraxinus mandshurica* var. *japonica*, *Quercus mongolica* var. *groseserrata*, etc., and the canopy height was 18-20 m. The basal area ranges from 13 (on the lower slope) to 56 (on the upper slope) m^2/ha in this forest area (Y. Nakagawa, unpublished data). Although canopy vegetation covers the entire northern watershed, none is observed in the part of the riparian wetland immediately above the outlet of the southern watershed (0.3 ha) (Figure 3).

Table 1. Climatic condition of the study site

Year	Mean annual temperature	Max. monthly Temp.		Min. monthly Temp.		Annual precipitation	Snowfall	Max. snow depth	
		°C				mm		cm	
2001	5.2	17.9	Aug.	-13.1	Feb.	1090	264	61	Feb.
2002	7.1	18.2	Jul.	-5.6	Jan.	1025	217	88	Feb.
2003	7.1	18.2	Aug.	-9.1	Jan.	1089	192	75	Mar.
2004	7.8	20.8	Aug.	-4.7	Jan.	976	359	135	Jan.
mean	6.8	—		—		1045	258	—	

Snowfall was regarded as the precipitation between Dec. and Mar. Month was when Max. and Min. values were observed.

Table 2. Properties of the two watersheds

Name	Location		Area (ha)	Riparian area (ha)*	Altitude (m)	Stream length (m)
Northern watershed	43°23′N	144°39′E	20.8	0.8	85-140	415
Southern watershed	43°20′N	144°37′E	57.3	1.9	35-145	260

* Riparian area is the flatarea around the foot of the slope.

While understory vegetation in both watersheds is mainly dwarf bamboo, *Sasa nipponica*, about 1-1.5 m tall, that in the riparian zone is *Carex spp.*, other herbaceous grass species, and *Urtica laetevirens* Maxim. of about 1-1.5 m tall.

Water Sampling

The stream waters were sampled in polyethylene bottles from baseflow every one or two months during the study period. The number of sampling points differed with the watershed, and stream length (Figure 2). The southern watershed had three points, i.e., downstream (S-STd), midstream (-STm) and upstream (-STu). The northern watershed had four points, i.e., downstream (N-STd), midstream (-STm), upstream (-STu) and upper spring (-STusp). In each watershed, the sampling points listed above were located every 120-150 m. The -STd in both watersheds was located at the outlet of the watershed, and S-STu in the southern watersheds and N-STusp in the northern watershed were located where perennial stream flow was apparent about 25 m downstream from the spring origin, and was assumed to be representative of the spring water for each watershed. Stream water sampling in the two watersheds was always conducted within one or two days of each other, except for N-STusp in the northern watershed, where the sampling began in November 2001. During the snow-melting and the soil-thawing period from March to May 2002, stream waters were sampled biweekly to evaluate the change in stream chemistry during the spring runoff period. While sampling the stream waters, the water temperature was also measured at each sampling point.

Soil water at a 30 cm depth and groundwater were also sampled at the bank alongside of the upstream (-STu) and midstream (-STm) sites in each watershed once every one or two months from August 2001 to 2002. At each site, soil and ground water were sampled at several points along a transect across the stream of the watershed (boxes and closed circles in Figure 4), the first point being 50 cm from the stream channel (near-stream sampling point). Slope soil waters were sampled 2 m above the foot of the slope. Numbers and locations of the sampling points varied with the width of the riparian zone (2-4 points). Soil water was collected using tension lysimeters with ceramic porous cups installed at a 30 cm depth, except for those near the stream. At each point described above, three replications of soil water were collected and were combined into one sample. Each tension lysimeter was evacuated using a 50 ml syringe without a filter one or two days before the collection. Ground water was collected from the wells constructed with 5 cm diameter PVC pipes screened from the soil surface to 80 cm or 100 cm depth in the southern watershed, and to 60 cm in the northern watershed. At the near-stream sampling point (at 50 cm away from the stream channel), ground water was also sampled using 5 cm diameter PVC pipes screened to 60-80 cm depth (hereafter called 70 cm depth) in the southern watershed, and to 40-60 cm depth (hereafter called 50 cm depth) in the northern watershed. In the northern watershed, the pipes for ground water sampling near the stream channel could not be installed to the deeper soil layer because of the presence of a stony and tight layer. Ground water was collected one or two days after the evacuation of the wells with a hand-pump. For each watershed, soil water and ground water were sampled on the same day when a series of the stream waters was sampled. Polyethylene bottles filled with stream water and groundwater contained few air bubbles to minimize the effect of degassing.

Between January and May 2002, tension lysimeters could not be used because of snow and soil freezing. Near the stream, subsurface soil at a 30 cm depth was saturated, and the water at this depth was collected using short-type PVC pipes (5 cm diameter) screened to 20-40 cm depth (hereafter called 30 cm depth). Although this water was the shallow ground water that had risen within 30 cm of the ground surface, we regarded these samples as the soil water.

Chemical Analyses

Water samples were brought back to the laboratory, and pH, EC, ORP (Oxidation-Reduction Potential) and pCO_2 (ORP and pCO_2 for 2001-2002) were measured immediately after arrival with glass electrode for pH and ORP, a meter for EC and membrane-glass electrode for pCO_2 (TOA DKK, CGP-1). Alkalinity was also measured only on stream water by the titration of 0.02N H_2SO_4 to pH 4.8 for 2001-2002 (The Japan Society for Analytical Chemistry, 1994).

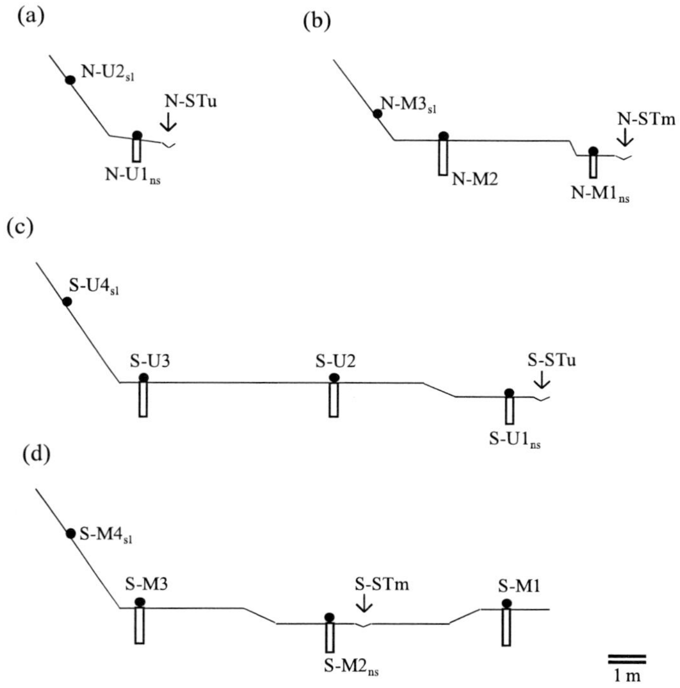

Figure 4. Sampling points of soil and ground water along the transect of the watershed. (a) Upstream and (b) midstream in the northern watershed, (c) upstream and (d) midstream in the southern watershed. Closed squares shows sampling point of the soil water at a 30 cm depth. Arrows show stream channels. The subscript "ns" denotes the near-stream sampling point (at 50 cm away from the stream channel). The subscript "sl" denotes the sampling point of the soil waters on the slope. Vertical boxes denote the approximate depths of sampling wells.

The soil water sampled by the tension lysimeter was not measured for pCO_2 because of a small amount. After the pH measurements, the water samples were passed through GF/F filters (Whatman Co. Ltd.) and were stored under 3 degrees centigrade until further chemical analyses were performed. Ion (Na^+, K^+, and Mg^{2+}, Ca^{2+}, NH_4^+, Cl^-, NO_3^- and SO_4^{2-}) concentrations were measured with ion chromatography (DIONEX Co. Ltd., DX-500) after passage through the 0.20 μm membrane filter (Toyo Roshi Kaisya Co. Ltd.). The concentration of dissolved organic carbon (DOC) was measured with a total organic carbon analyzer (Shimadzu Inc., TOC-5000). Dissolved iron (Fe), aluminum (Al), silica (Si) and phosphorous (P) concentrations were measured by inductively coupled plasma optical emission spectrometry (Japan JARRELL-ASH Co. Ltd.). The concentrations of ions and dissolved solutes and pCO_2 were expressed in μmolc/L, μmol/L, and %, respectively.

Estimation of Stream pH

Stream pH was calculated following the equation by Ohte et al. (1995),

$$[H^+] = , \tag{1}$$

to confirm the difference between measured and calculated pH. In this equation, ALK denotes alkalinity. This equation was obtained from the following equations:

$$ALK = [HCO_3^-] - [H^+] + [OH^-] \tag{2}$$

$$[H^+][HCO_3^-]/pCO_2 = 10^k \tag{3}$$

We used the alkalinity titrated and k at 15 degrees centigrade (-7.74, Stumm and Morgan 1981). For equation (1), lowered alkalinity and elevated pCO_2 enhance H^+ concentration. However, this equation (1) did not take into account the contribution of organic acid anions. When alkalinity contains organic acid anions, ALK should be,

$$ALK = [HCO_3^-] + OAA - [H^+] + [OH^-]. \tag{4}$$

OAA denotes the total concentration of the organic acid anions assuming the OAA carry a charge of -1. When alkalinity depends on a significant portion of organic acid anions, equation (1) underestimates H^+ concentration and therefore leads to an overestimation of pH.

Measurement of Stream Water Flow

On 24 October, 2002, during base-flow, stream flow was measured with a flow meter having a propeller (KENEK Co., Ltd., VR-101), and then was calculated under consideration of the stream cross section (Arai, 1994).

Statistical Analyses

The change in water chemistry along the stream was examined by analysis of variance followed by Scheffe's multiple comparison ($p<0.05$).

For NH_4^+, NO_3^-, Fe and Al concentrations in the stream water, soil water and groundwater, multiple regression analyses were performed to examine the relation of pH, ORP and DOC concentration on them. Variables were selected by the statistical significance ($p<0.05$) on the stepwise-method.

Variables were transformed to normal distribution if necessary.

RESULTS

Stream Water Flow

Stream water flow of the upstream (S-Stu), midstream (S-STm) and downstream (S-STd) on 24 Oct. 2002 in the southern watershed was 0.4, 6.9, and 9.7 L/s, respectively. In the northern watershed, stream water flow also increased from the spring to the downstream: 0.8 (at N-STusp), 7.2 (at N-STu), 10.9 (at N-STm) and 15.7 (at N-STd) L/s.

Seasonal Change of Stream Water Temperature

Seasonal changes in stream water temperatures in the northern watershed were similar among the sampling points (Figure 5a). On the other hand, those in the southern watershed differed with the sampling points; the largest variation was found at the upstream, the smallest variation at the midstream, and medium variation at the downstream. Variation patterns of the stream temperature in the southern watershed were similar to that at the downstream in the northern watershed.

Spatial Variety of Stream Water Chemistry

Stream pH was increased downstream in both watersheds (Figure 5, Table 3). Most ion concentrations decreased downstream except for NH_4^+ in both watersheds and Cl^- in the southern watershed. For the ions exhibiting downstream decrease in the southern watershed (measured ions except for NH_4^+ and Cl^-), concentrations significantly decreased from S-STu to S-STm, however, did not significantly change from S-STm to S-STd (Table 3).

The decreasing patterns of ion concentrations in the northern watershed varied among ions. Stream NH_4^+ concentrations in both watersheds did not significantly change downstream. Stream Cl^- concentration in the southern watershed significantly increased from S-STu to S-STm, however, did not significantly change from S-STm to S-STd.

Table 3. Results of ANOVAs and the following multiple comparisons (Scheffe's method) for the stream water chemistry. different small characters denote the statistical significance (p<0.05)

Solute	Position	Southern Watershed				Northern Watershed			
		Mean	SD	*F*	*p*	Mean	SD	*F*	*p*
pH	-STusp			80.8	<0.001	6.76	0.07 c	282.0	<0.001
	-STu	6.93	0.10 c			7.11	0.11 b		
	-STm	7.09	0.06 b			7.51	0.13 a		
	-STd	7.22	0.09 a			7.52	0.11 a		
Na^+ (μmolc/L)	-STusp			376.6	<0.001	343.3	12.8 ab	5.3	<0.001
	-STu	340.9	20.3 a			346.9	17.3 a		
	-STm	249.8	9.9 b			337.3	17.0 ab		
	-STd	247.5	10.0 b			330.6	16.3 b		
K^+ (μmolc/L)	-STusp			445.1	<0.001	55.2	2.6 a	8.9	<0.001
	-STu	78.0	5.6 a			53.8	7.0 a		
	-STm	49.5	2.4 b			51.3	6.0 ab		
	-STd	49.7	3.4 b			48.0	4.9 b		
Mg^{2+} (μmolc/L)	-STusp			1146.6	<0.001	222.9	11.7 a	14.3	<0.001
	-STu	253.7	18.8 a			228.5	6.4 a		
	-STm	123.1	4.7 b			215.5	8.4 b		
	-STd	121.5	5.6 b			215.0	8.5 b		
Ca^{2+} (μmolc/L)	-STusp			745.7	<0.001	643.6	35.0 a	32.8	<0.001
	-STu	629.7	63.2 a			649.8	30.6 a		
	-STm	269.2	18.7 b			604.4	36.1 b		
	-STd	260.6	21.9 b			567.7	35.2 c		
NH_4^+ (μmolc/L)	-STusp			2.2	0.114	0.00	0.00	0.8	0.499
	-STu	4.58	9.69			0.45	1.80		
	-STm	2.28	5.61			0.08	0.43		
	-STd	0.84	1.74			0.37	1.62		
Cl^- (μmolc/L)	-STusp			14.6	<0.001	181.1	12.8 a	65.1	<0.001
	-STu	63.8	7.8 b			169.6	13.2 b		
	-STm	71.9	4.8 a			150.3	12.4 c		
	-STd	72.0	6.1 a			137.2	10.8 d		

Solute	Position	Southern Watershed Mean	SD	F	p	Northern Watershed Mean	SD	F	p
	-STusp					142.8	18.0 a		
NO_3^-	-STu	49.8	17.4 a	111.3	<0.001	135.2	18.2 a	59.8	<0.001
(μmolc/L)	-STm	14.8	4.3 b			105.5	16.1 b		
	-STd	11.7	2.3 b			90.1	12.6 c		
	-STusp					29.2	1.4 a		
SO_4^-	-STu	50.6	7.1 a	79.1	<0.001	25.8	3.0 b	21.7	<0.001
(μmolc/L)	-STm	37.9	3.1 b			24.9	2.5 b		
	-STd	36.2	1.7 b			24.2	2.2 b		
	-STusp					67.6	14.1		
DOC	-STu	98.6	23.9 a	17.9	<0.001	56.9	14.0	3.1	0.030
(μmol/L)	-STm	67.7	14.8 b			68.4	19.2		
	-STd	86.7	17.6 ab			60.9	16.0		
	-STusp					0.10	0.10 a		
Fe	-STu	0.09	0.11 c	189.2	<0.001	0.04	0.05 b	3.7	0.015
(μmol/L)	-STm	0.91	0.19 b			0.06	0.05 ab		
	-STd	2.49	0.77 a			0.09	0.07 ab		
	-STusp					0.68	0.40		
Al	-STu	0.74	0.52 a	5.1	0.008	0.71	0.47	0.1	0.940
(μmol/L)	-STm	0.41	0.34 b			0.75	0.47		
	-STd	0.45	0.37 b			0.68	0.37		
	-STusp					0.51	0.43		
P	-STu	0.47	0.48	3.7	0.030	0.71	0.59	1.4	0.238
(μmol/L)	-STm	0.78	0.55			0.58	0.51		
	-STd	0.46	0.43			0.79	0.60		
	-STusp					990.6	50.6		
Si	-STu	892.1	70.5 b	3.2	0.045	1003.7	48.1	1.0	0.409
(μmol/L)	-STm	934.0	51.6 a			981.9	54.0		
	-STd	910.7	59.1 ab			985.6	48.4		

DOC concentration in the southern watershed significantly decreased from S-STu to S-STm, and then increased to the S-STd although not to the S-STu level. That in the northern watershed did not significantly change downstream.

Fe concentration in the southern watershed significantly increased downstream. That in the northern watershed showed slightly and significantly higher value in N-STusp than in other stream points. Al concentration in the southern watershed significantly decreased from S-STu to S-STm, and then did not significantly change from S-STm to S-STd. P concentrations in both watersheds did not significantly change downstream. Si concentration in the southern watershed significantly increased from S-STu to S-STm, and then decreased to the S-STd although not to the S-STu level, exhibiting the pattern opposite to the DOC concentration. Si concentrations in the northern watersheds did not significantly change downstream.

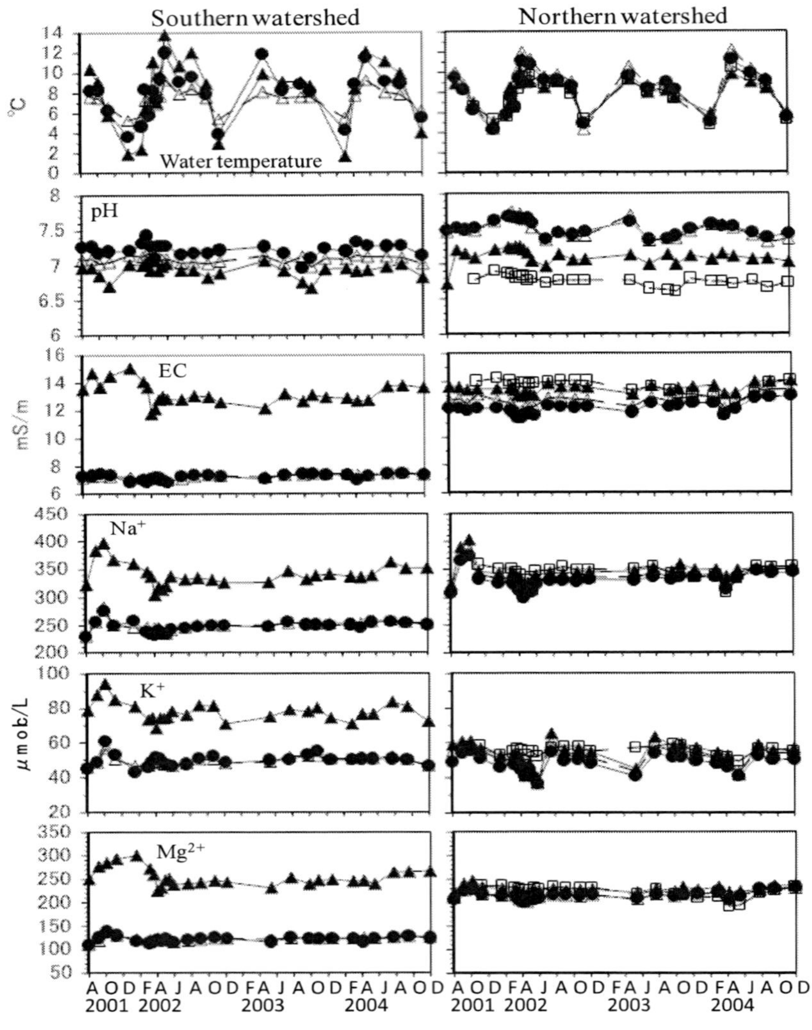

Figure 5. (Continued).

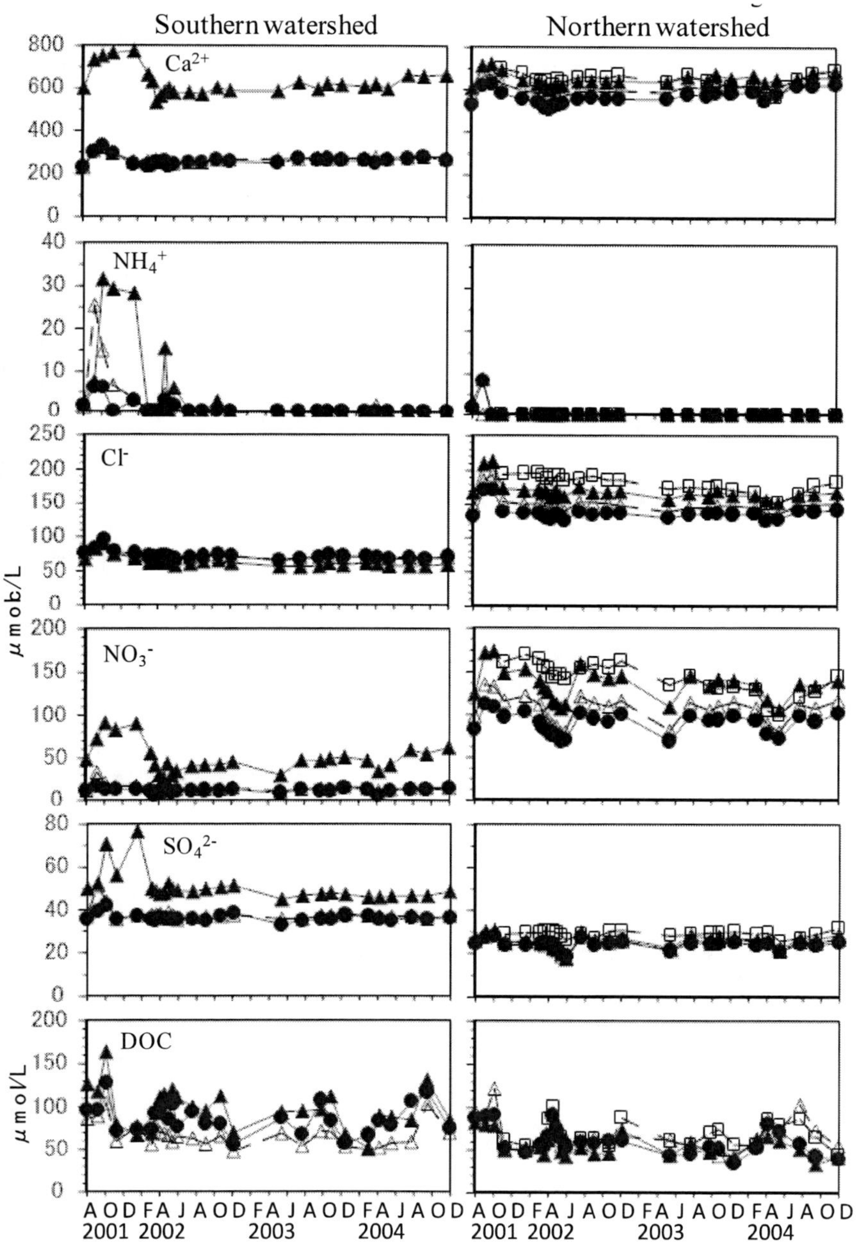
Southern watershed
Northern watershed
Ca2+
NH4+
Cl-
NO3-
SO42-
DOC
μmol/L
μmol/L
A O D F A J A O D F A J A O D F A J A O D
2001 2002 2003 2004

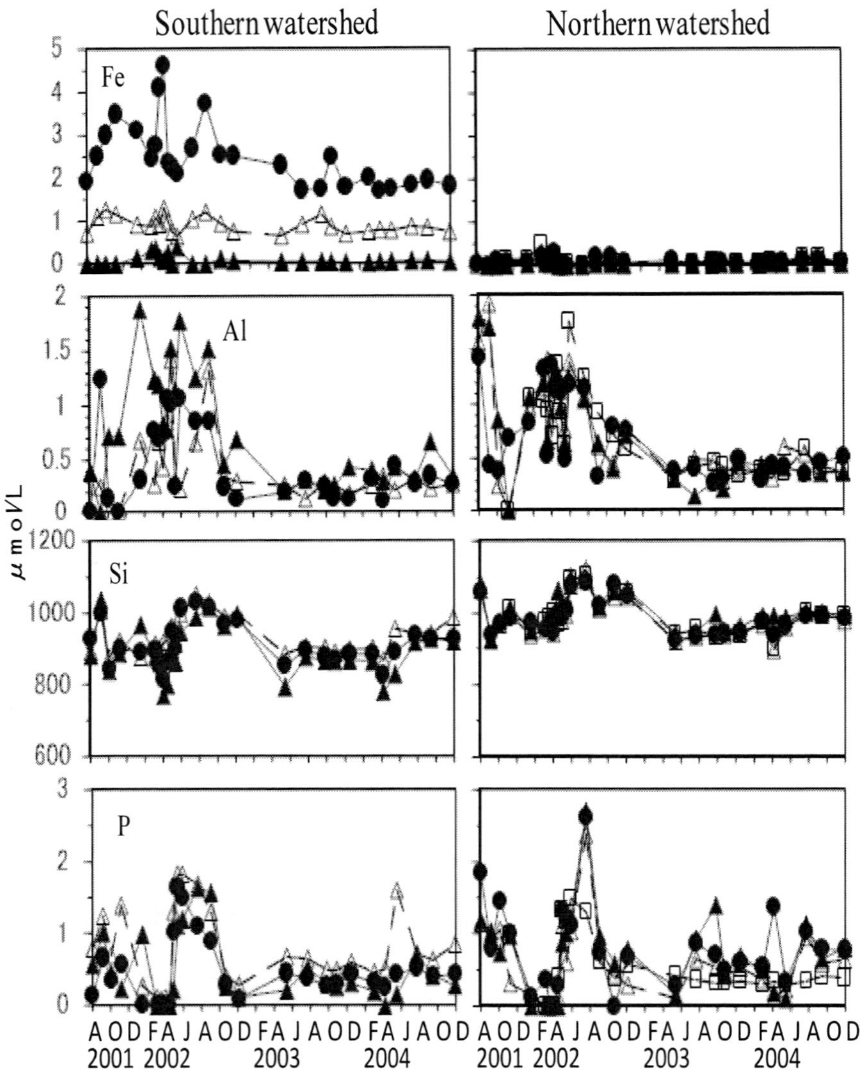

Figure 5. Spatial and Temporal variations of the stream water chemistry. □: upper spring (-STusp), ▲: upstream (-STu), △: midstream (-STm), ●: downstream (-STd).

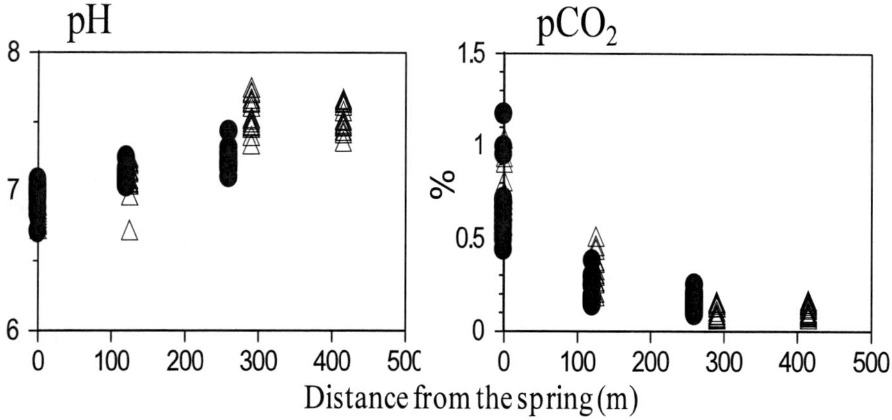

Figure 6. Variations of the stream pH and pCO2 along the distance from the spring. ●: southern watershed, ∆: northern watershed.

Relation of the pCO_2 to the Stream pH

Stream water originates from the spring. As shown in Figure 5, stream pH increased downstream (Figure 6). Contrastingly, pCO_2 decreased downstream, and moreover converged. In the northern watershed having longer stream, pCO2 decreased and converged more than in the southern watershed.

Figure 7a shows the close relationships between pCO_2 and pH in both watersheds, although the pH values in the lower pCO_2 range (< 0.5 %) were lower in the southern watershed than in the northern watershed. Although not shown, other chemicals did not show a close correlation with pH as pCO_2. Alkalinity (mean ± SD μmolc/L) at the downstream in the southern (S-STd) and northern watershed (N-STd) in 2001-2002 were 560.9 ± 22.6 and 895.7 ± 23.6 μmolc/L, respectively. Furthermore, stream pH calculated using alkalinity and pCO_2 following equation (1) well agreed with the pH measured (Figure 7b).

pH and pCO_2 in the Soil and Ground Water

In both watersheds, pH in the soil and ground waters were mostly lower than those in the stream waters, although the pH of the ground waters near the upstream in the southern watershed (S-U1 group) was similar to the upstream pH (Figure 8). Conversely, most pCO_2 of the ground water was higher than that in the stream waters, although the pCO_2 in the ground waters near the upstream in the southern watershed was slightly lower than the upstream pCO_2 (Figure 9). In the southern watershed, ground waters near the stream, S-U1$_{ns}$ group and S-M2$_{ns}$ groups, had a lower pCO_2 than the other ground waters, at S-U2, -U3 and S-M1, -M3, respectively. Moreover, pCO_2 of ground waters at a 70 cm depth near the upstream (S-U1-70) was lower than that at the upstream and at same point, S-U1$_{ns}$ and -U1$_{ns}$ -30 in the southern watershed. Near the midstream in the southern watershed, similarly, pCO_2 of ground water at 70 cm depth (S-M2$_{ns}$ -70) was lower than those at S-M2$_{ns}$ and S-M2$_{ns}$ -30. As a whole, pCO_2 in the ground waters was higher in the southern watershed than in the northern watershed, although pH in the soil water and ground water were similar in both watersheds (Figure 8).

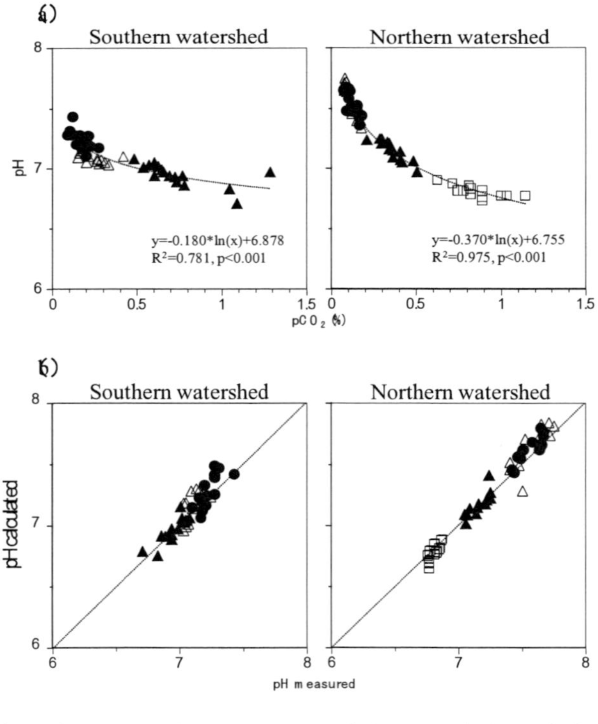

Figure 7. Relation of stream pCO_2 to stream pH (2001-2002). (a) Relationships between pH and pCO_2 in the stream water. (b) Relationship between pH measured and those calculated using alkalinity and pCO_2 (equation 1). Dotted lines in (a) denote regression lines, and those in (b) denote 1:1. Legend is same as Figure 5.

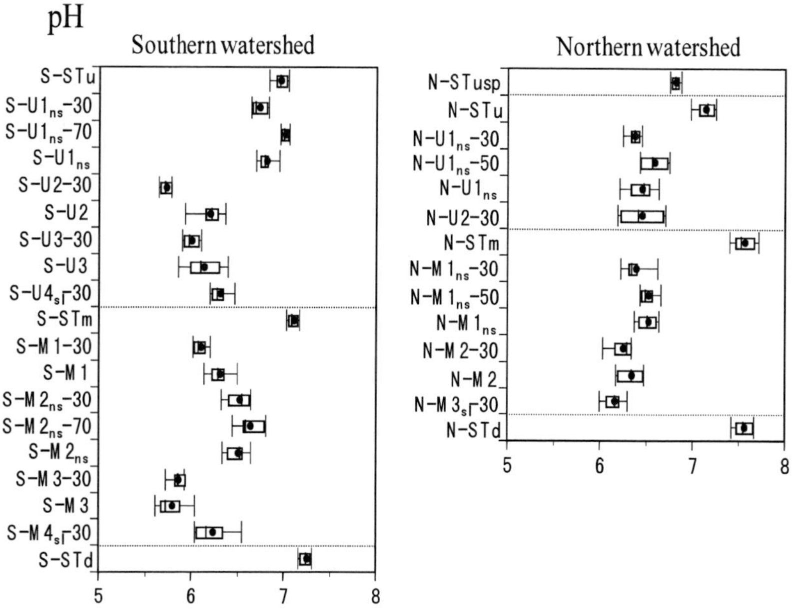

Figure 8. Distributions of pH in the stream, soil and ground waters (2001-2002). Figure after the second hyphen denotes the depth where water was sampled. The water of no figure, except for the stream water, denotes ground water sampled from the well fully screened. 10, 25, 50, 75, 90 percentile of the concentration distribute. Closed circles denote mean concentrations.

NO_3^- and Fe Concentrations in the Soil and Ground Water

In the southern watershed, NO_3^- concentrations in the soil and ground water adjacent to upstream (0.6-66 μmolc/L, S-U1$_{ns}$ ~ -U4$_{sl}$ groups) were lower than those at the upstream (26-91μmolc/L, S-STu) (Figure 10). Highest NO_3^- concentrations in the soil and ground water (S-U3 and -U4$_{sl}$ groups) were observed at the slope and the foot of the slope. NO_3^- concentrations in the soil and ground water adjacent to midstream (< 3.2 μmolc/L, S-M1$_{ns}$ ~ -M4$_{sl}$ groups) were much lower than those adjacent to the upstream and in the three stream water sampling locations (5.8-77 μmolc/L, S-STu, -STm and -STd).

In the northern watershed, NO_3^- concentrations in the soil and ground water (n.d.-90 μmol/L) were lower than those in stream waters (82-166 μmolc/L, i.e. N-STusp, -STu, -STm and -STd), although high NO_3^-

concentrations were contained in 50 cm depth ground water near the upstream (ca. 110 μmolc/L, N-U1_{ns}-50).

In the southern watershed, Fe concentrations in the soil water and ground water near the upstream (< 9.1 μmol/L, S-U1_{ns}-30 and SU1_{ns}) were higher than that at the upstream (< 0.4 μmol/L, S-STu), whereas those in the ground water at a 70 cm depth (< 0.2 μmol/L, S-U1_{ns} -70) was similar to the value at the upstream (Figure 11). Other soil and ground water (S-U2 and S-U3 groups) contained higher Fe concentrations (n.d.-113 μmol/L) than the upstream. Fe concentrations in soil and ground water adjacent to the midstream (< 113 μmol/L, S-M1 ~ -M4_{sl} groups) were higher than that at the midstream (0.6-1.3 μmol/L, S-STm). Low Fe concentrations were found in some soil water (S-U4_{sl} and -M4_{sl}).

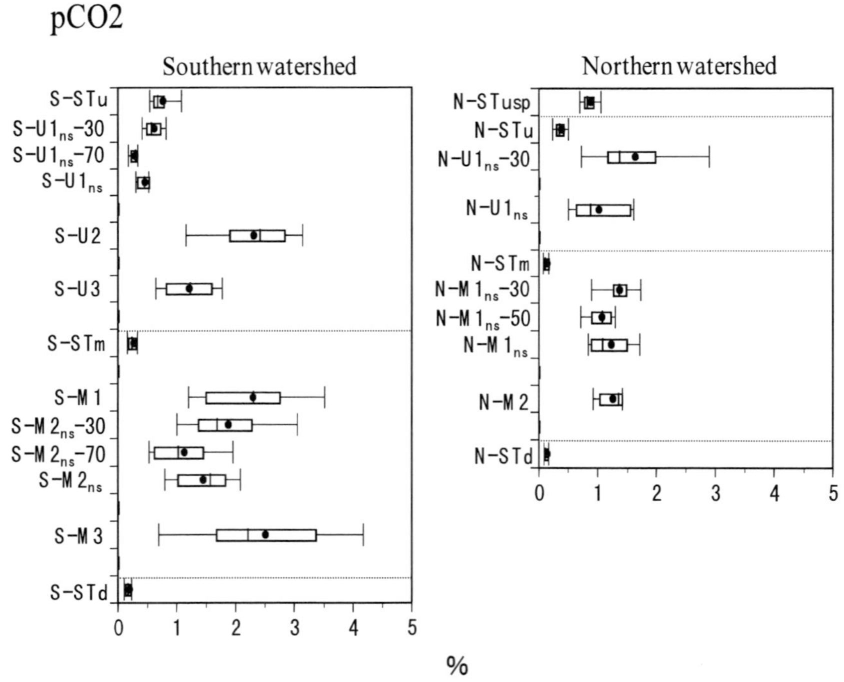

Figure 9. Distributions of pCO_2 in the stream, soil and ground waters (2001-2002). Figure after the second hyphen denotes the depth where water was sampled. The water of no figure, except for the stream water, denotes ground water sampled from the well fully screened. 10, 25, 50, 75, 90 percentile of the concentration distribute. Closed circles denote mean concentrations.

In the northern watershed, Fe concentrations in the soil and ground water (< 340 μmol/L) were higher than those in stream waters (< 0.5 μmol/L), although low Fe concentrations (< 0.5 μmol/L) were sometimes found in some soil waters (N-U2$_{sl}$-30 and -M3$_{sl}$-30). As a whole, the Fe concentrations in the ground water were lower in the northern watershed than in the southern watershed except for a very high concentration in the ground water at a 50 cm depth near the upstream in the northern watershed (> 200 μmol/L, N-U1$_{ns}$-50).

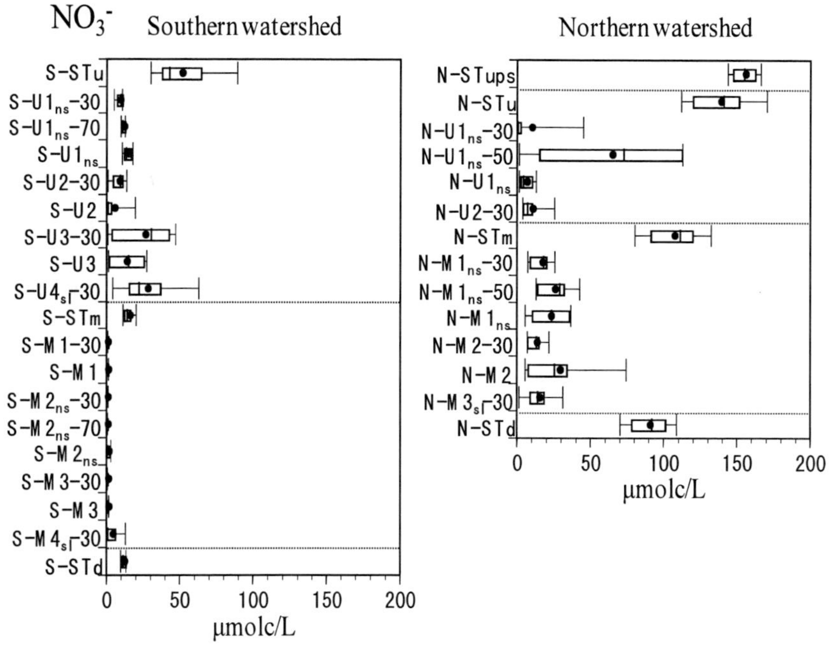

Figure 10. Distributions of NO_3^- concentrations in the stream, soil and ground waters (2001-2002). Figure after the second hyphen denotes the depth where water was sampled. The water of no figure, except for the stream water, denotes ground water sampled from the well fully screened. 10, 25, 50, 75, 90 percentiles of the concentration distribute. Closed circles denote mean concentrations.

DOC Concentration in the Soil and Ground Water

In the southern watershed, DOC concentrations in the soil and ground water near the upstream (42-188 μmol/L, S-U1_{ns}, and -U1_{ns} -30) were similar to that at upstream (70-164 μmol/L, S-STu), whereas that of 70 cm ground water (14-46 μmol/L, S-U1_{ns} -70) was lower than those in S-U1_{ns}, S-U1_{ns} -30 and upstream (Figure 12).

On the other hand, the other soil and ground water (S-U2, -U3 and -U4_{sl} groups) exhibited higher DOC concentration (> 100 μmol/L) than the upstream. DOC concentrations in the soil and ground water adjacent to the midstream (109-2652 μmol/L, S-M1 ~ -M4_{sl} groups) were higher than those at the midstream (48-105 μmol/L, S-STm), although the soil and ground water of S-M2_{ns} group exhibited relatively low DOC concentrations (109-431 μmol/L). In the ground water at a 70 cm depth (S-M2_{ns}-70), DOC concentrations (109-357 μmol/L) were still lower than the other shallow soil and ground water adjacent to the midstream.

In the northern watershed, DOC concentrations in the soil and ground water adjacent to the upstream (> 100 μmol/L) were higher than that at the upstream (43-74 μmol/L, N-STu). DOC concentrations in the soil and ground water near the midstream (53-249 μmol/L, N-M1_{ns} group) were slightly higher than that at the midstream (56-122 μmol/L, N-STm), whereas the other soil and ground water (N-M2 and N-M3_{sl} groups) exhibited apparently higher DOC concentrations (254-2691 μmol/L) than the midstream.

Multiple Regression Analyses for NH_4^+, NO_3^-, Fe and Al Concentrations

For the relations of pH, ORP and DOC concentration to the NH_4^+, NO_3^-, Fe and Al concentrations, the phenomenon common to both watersheds were in the following (Table 4).

1. NH_4^+ concentration had an inclination to be higher in the water having lower pH.
2. NO_3^- concentration had an inclination to be higher in the water having higher pH.
3. Fe and Al concentrations had inclinations to be higher in the water having higher DOC concentrations.

4. Al concentration had an inclination to be higher in the water having higher ORP.

Fe concentration had an inverse inclination in the respective watersheds.

In the southern watershed, otherwise, NO_3^- concentration had an inclination to be higher in the water having higher ORP, and Al concentration had an inclination to be higher in the water having lower pH.

In the northern watershed, NH_4^+ concentration had an inclination to be higher in the water having lower ORP and higher DOC concentration. Moreover, NO_3^- concentration had an inclination to be higher in the water having lower DOC concentration, and Fe concentration had an inclination to be higher in the water having lower pH and higher ORP.

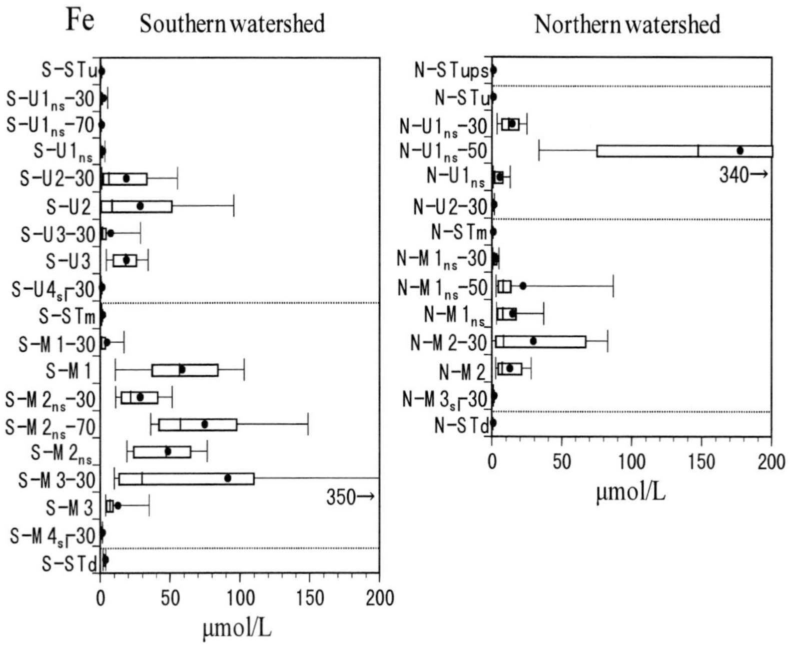

Figure 11. Distributions of Fe concentrations in the stream, soil and ground waters (2001-2002). Figure after the second hyphen denotes the depth where water was sampled. 10, 25, 50, 75, 90 percentiles of the concentration distribute. Closed circle denotes mean concentration. High concentrations > 300 μmol/L were found three times at N-U1$_{ns}$-50 in the northern watershed and once at S-M3-30 in the southern watershed.

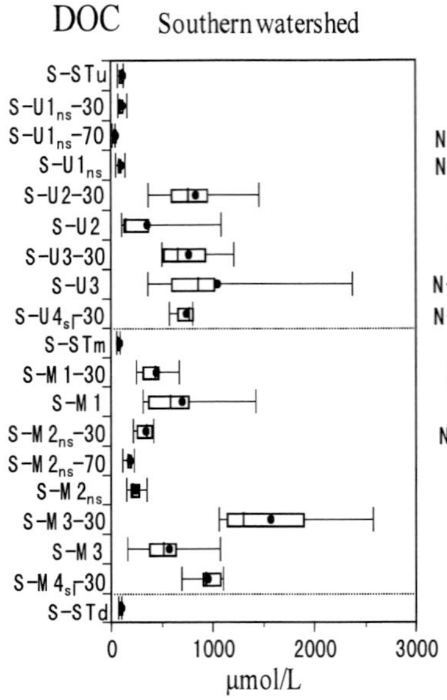

Figure 12. Distributions of DOC concentrations in the stream, soil and ground waters (2001-2002). Figure after the second hyphen denotes the depth where water was sampled. 10, 25, 50, 75, 90 percentiles of the concentration distribute. Closed circles denote mean concentrations.

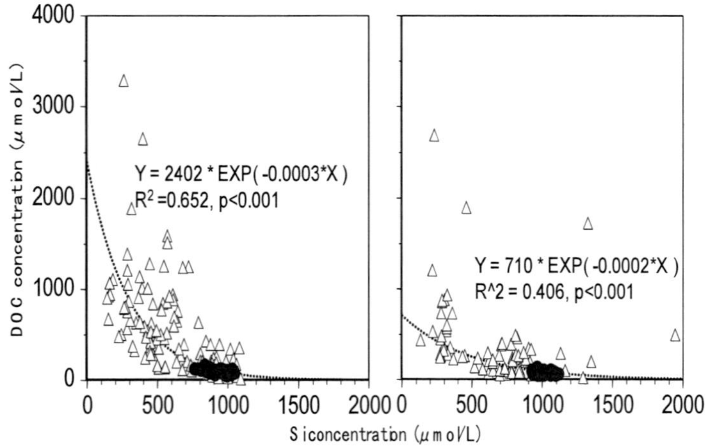

Figure 13. Relationships between Si and DOC concentrations (2001-2002). Dotted lines denote regression lines. ●: stream water, Δ: soil water and groundwater.

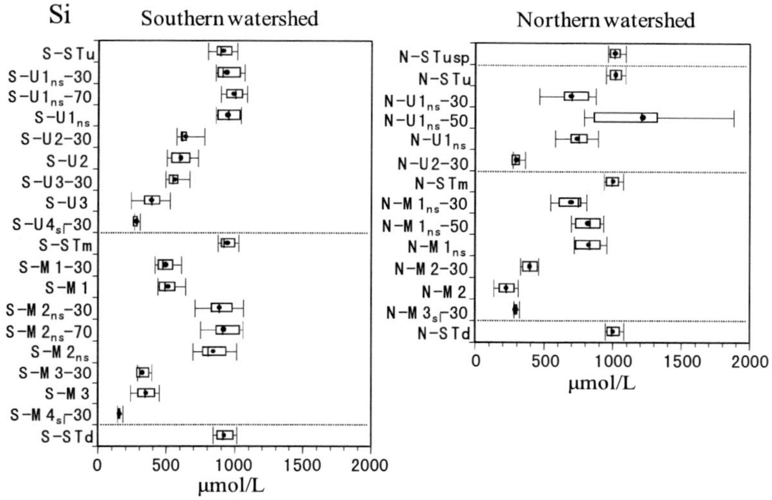

Figure 14. Distributions of Si concentrations in the stream, soil and ground waters (2001-2002). Figure after the second hyphen denotes the depth where water was sampled. 10, 25, 50, 75, 90 percentiles of the concentration distribute. Closed circles denote mean concentrations.

Relationship between Si and DOC Concentrations in the Soil and Ground Water

The relationship between Si and DOC concentrations in the stream water, soil water and ground water showed significantly negative correlation in the both watershed (Figure 13).

In the southern watershed, Si concentrations in the soil and ground water near the stream channel (S-U1_{ns} and -M2_{ns} groups) were similar to those in the stream water (Figure 14). The deeper ground water near the stream channel (S-U1_{ns}-70 and -M2_{ns}-70) in the southern watershed, where slightly lower DOC concentrations were shown (Figure 12), had slightly higher Si concentrations than the other ground water. Si concentrations in the other soil and ground water, in the southern watershed, were lower than those in the stream water (Figure 14).

In the northern watershed, Si concentrations in the soil and ground water were lower than those in the stream water, although some of the deep ground water (N-U1_{ns}-50) had the Si concentrations higher than the stream water.

In both watersheds, Si concentrations in the soil and ground water increased from the slope to the stream channel, although the northern watershed had the sampling points fewer than the southern watershed.

Relationship between Na^+ and Cl^- Concentrations in the Soil and Ground Water

The relationship between Na^+ and Cl^- concentrations in the stream water, soil water and ground water showed significantly positive correlation in the northern watershed, but no significant correlation in the southern watershed (Figure 15).

In the southern watershed, Na^+ concentrations in the soil and ground water adjacent to the upstream were lower than that in the upstream (S-STu), although some exceptions (Figure 16). Na^+ concentrations in the soil and ground water near the midstream (S-M1 and S-$M2_{ns}$ groups) in the southern watershed were slightly higher than that at the midstream (S-STm). Among the soil and ground water, Na^+ concentrations at the foot of the slopes and at the slopes (S-U3 group, -$U4_{sl}$-30, -M3 group and -$M4_{sl}$-30) were lower than in the other soil and ground water.

In the northern watershed, Na^+ concentrations in the soil and ground water were lower than those in the stream water, although some exceptions. Similarly to the southern watershed, Na^+ concentrations at the foot of the slopes and at the slopes (N-U2-30, -U2 group and -$M3_{sl}$-30) in the northern watershed were lower than those in the other soil and ground water.

Table 4. Results of multiple regression analyses for some solutes. t value and probability of the partial regression coefficiency, and F value and probability of the Anova

Solute	Southern Wartershed										Northern Wartershed									
	pH		ORP		DOC		Constant		ANOVA		pH		ORP		DOC		Constant		ANOVA	
	t	*p*	*t*	*p*	*t*	*p*	*t*	*p*	*F*	*p*	*t*	*p*	*t*	*p*	*t*	*p*	*t*	*p*	*F*	*p*
NH_4^+	-3.7	<0.001					4.6	<0.001	14.0	<0.001	-2.6	0.011	-3.8	<0.001	3.7	<0.001	2.9	0.005	21.8	<0.001
NO_3^-	6.9	<0.001	5.7	<0.001			-8.0	<0.001	51.5	<0.001	20.6	<0.001			-10.3	<0.001			998.1	<0.001
Fe			-8.2	<0.001	13.2	<0.001			219.1	<0.001	-4.0	<0.001	3.0	0.004	2.7	0.007			31.2	<0.001
Al	-8.5	0.022	2.8	0.007	20.4	<0.001			523.1	<0.001			2.7	0.007	6.2	<0.001	-3.7	<0.001	23.0	<0.001

For Cl⁻, concentrations in the soil and ground water in the southern watershed were usually higher than those in the stream water, although some exceptions (S-U4$_{sl}$-30, S-M4$_{sl}$-30 and some occasions) (Figure 17). In the northern watershed, Cl⁻ concentrations in the soil and ground water were lower than those in the stream water, although some exceptions.

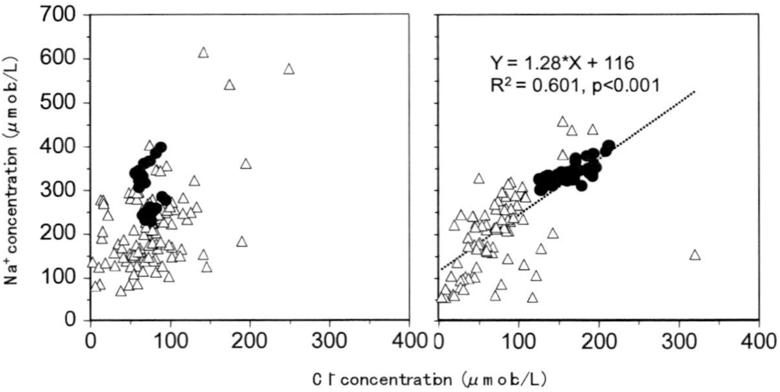

Figure 15. Relationships between Na^+ and Cl^- concentrations (2001-2002). Dotted line denotes a regression line. ●: stream water, Δ: soil water and groundwater.

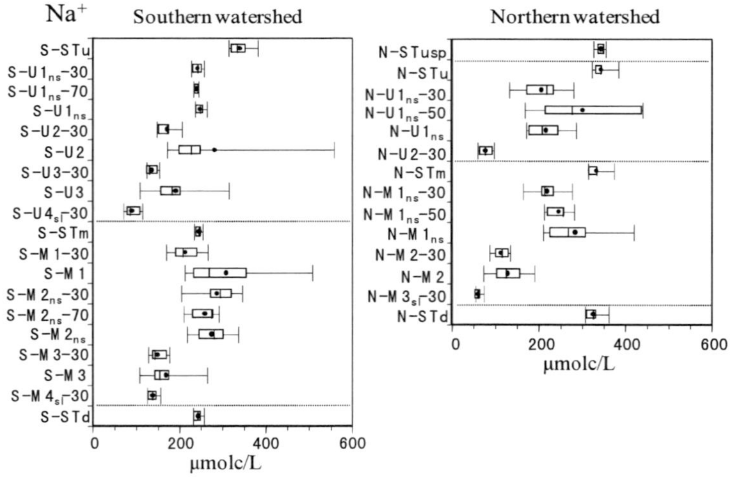

Figure 16. Distributions of Na^+ concentrations in the stream, soil and ground waters (2001-2002). Figure after the second hyphen denotes the depth where water was sampled. 10, 25, 50, 75, 90 percentiles of the concentration distribute. Closed circles denote mean concentrations.

Discussion

Difference in Hydrological Regime between Two Watersheds

In both watersheds, stream water flow increased further downstream, because the surrounding soil and ground water flowed into the stream channel. However, longitudinal patterns of solute concentrations were different between both watersheds (Figure 5). These differences in solute dynamics would reflect the different influences of the soil and ground water around the stream channel on the stream water chemistry.

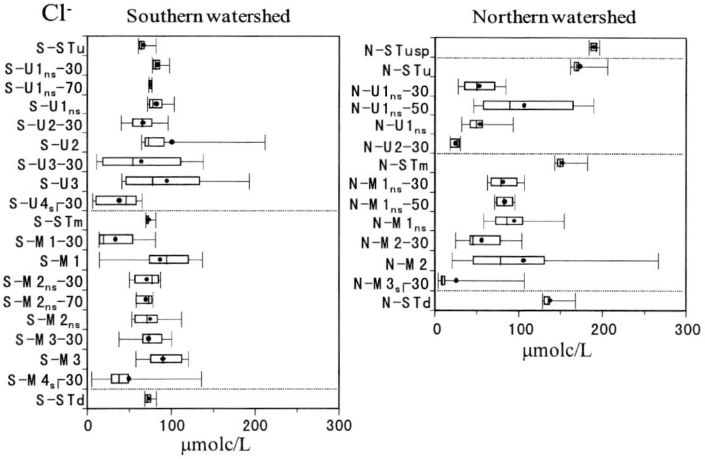

Figure 17. Distributions of Cl^- concentrations in the stream, soil and ground waters (2001-2002). Figure after the second hyphen denotes the depth where water was sampled. 10, 25, 50, 75, 90 percentiles of the concentration distribute. Closed circles denote mean concentrations.

In the present study, the stream water flow at the outlet of the watershed was higher in the small watershed (i.e. northern watershed) than in the large watershed (i.e. southern watershed). The possible reasons for the disagreement between the stream flow and watershed area were 1) that effective water source area of the northern watershed would be larger than the area based on the surface topography and/or 2) that the southern watershed might have some effective pathways of water effluent other than the stream channel. Moreover, the difference in the pattern of stream water temperature further downstream between both watersheds suggests the different pathways of the groundwater inflow. In the southern watershed, the attenuated fluctuation in the water temperature from the upstream to midstream suggested the increasing contribution of relatively deep ground water to the stream water in this reach, because deep soil layer was less affected by air temperature variation than the surface soil layer. The amplified fluctuation in the stream temperature from the midstream to downstream suggested the decreasing contribution of deeper ground water. The similar fluctuations of the temperature among the steam points in the northern watershed might suggest a relatively constant contribution of both shallow and deep ground water to the stream water.

Effect of the pCO_2 on the Stream pH

The ground water had the high pCO_2 due to the CO_2 production by plant roots and soil microbes, and this high pCO_2 was reflected in higher pCO_2 and lower pH of the spring water (Jones and Mulholland 1998; Ohte et al. 1995). For both watersheds, the main process causing the spatial variation of stream pH was degassing of CO_2: lowering pCO_2 and causing an increase in pH downstream (Choi et al. 1998). Lower pH at the downstream in the southern watershed (S-STd) than in the northern watershed (N-STd) was of lower alkalinity and less complete degassing of CO_2 in the southern stream than in northern stream (Figure 6).

The soil and ground water having higher pCO_2 and lower pH flowed into the stream channel. However, pCO_2 in the groundwater would decrease to the level of stream pCO_2, as ground water approached the stream channel. This CO_2 decrease probably occurred mainly at the soil-stream interface, i.e., the soil just close to the stream channel, because pCO_2 in the ground waters 50 cm away from the stream channels did not decrease to the stream pCO_2 except for the spring water, i.e., S-STu in the southern watershed and N-STusp in the northern watershed (Figure 9). This complete degassing might be due to slow seepage of groundwater to stream relative to CO_2 degassing. The stream water at the upstream (S-STu) had pCO_2 slightly higher than the ground water of S-$U1_{ns}$ group (Figure 9), probably because the stream water at the upstream in the southern watershed, assumed to be the spring water, might be subject to somewhat higher pressure than the ground water of S-$U1_{ns}$ group.

Influence of the Riparian Zone on the Interrelation among NO_3^-, Fe and DOC Concentrations in the Stream Chemistry

In both watersheds, because the NO_3^- concentrations in the soil and ground water were lower than those in stream waters (Figure 10), the decrease of NO_3^- concentration further downstream would be caused mainly by the dilution with the surrounding soil and ground water having lower concentrations and/or removal at the riparian zone, which were probably due to denitrification and plant root uptake (Bowden, 1987; Hedin et al., 1998). Low NO_3^- concentrations in the ground water might be cause partially by the nitrification inhibition due to low pH resulting in high DOC and NH_4^+ concentrations (Hart et al. 1994) (Table 4). Moreover, unchanged NO_3^- concentration from S-STm to S-STd in the southern watershed (Table 3) suggested the nitrification at the soil-stream interface and/or in the stream channel.

On the other hand, the DOC and Fe concentrations in the ground water were higher than those in stream waters, although the some soil water had similar concentrations as those in stream waters (Figure 11 and 12). Under the water saturated and anaerobic conditions, dissolution of undecomposed matter increases DOC concentration (Dalva and Moore, 1991) and Fe is reduced and most of the dissolved iron would be stabilized by chelating organic carbon (Suzuki et al., 1992). Low pH in the ground water might induce the Fe dissolution, especially in the northern watershed (Table 4). The increase of Fe concentration further downstream in the southern watershed would be mainly by the inflow of the ground water having high Fe into the stream channel, although a similar increase of Fe in the northern watershed was not found (Figure 5 and Table 3). Fe was probably precipitated at the soil-stream interface (within 50 cm from the stream channel), where the soil water was not collected in the present study, and/or in the stream, especially in the northern watershed. Moreover, the DOC concentrations in either stream did not increase in spite of the inflow of the ground water having high DOC (Figure 12), suggesting a sink of DOC at the soil-stream interface and /or an in-stream consumption of DOC. Two possible processes can explain this DOC sink. There may have been sorption of DOC in the soil close to the stream channel (Nelson et al. 1993). Alternatively, there may have been metabolic consumption by microbes at the soil-stream interface and/or by microbes and benthos in the sediment, etc. (Ford and Naiman, 1989; Hedin et al., 1998). Furthermore, a similar metabolic consumption would occur as the in-stream process.

In the southern watershed, as a whole, the same DOC sink processes were important. However, only the same processes could not account for the longitudinal pattern of stream DOC concentration in the southern watershed. The soil and ground water near the upstream had DOC concentrations similar to the stream water at the upstream, representing the DOC concentration in the spring water. However, the deeper ground water at 70 cm (S-U1_{ns}-70 and -M2_{ns}-70) exhibited slightly lower DOC concentration than other groundwater at the same position (S-U1 and -M2 groups, respectively). Therefore, the significant decrease of DOC concentration from the upstream to midstream was due to the contribution of this deeper groundwater to this reach. The significant increase of the DOC concentration from the midstream to downstream was due to the disappearance of this contribution by the deep groundwater and some enrichment by the surrounding groundwater having high DOC. This transition of contribution by the different groundwater was according to the transition of the seasonal change in the stream water temperature (Figure 5).

Hedin et al. (1998) suggested that riparian wetland was under somewhat oxidizing condition because ground water mixes with oxidized stream water. Probably, the soil-stream interface in the present study was also under some oxidizing condition. Therefore, nitrification and Fe precipitation at the soil-stream interface might be reasonable. Furthermore, the consumption of DOC at the same position also might be possible through the metabolic activity by soil microbes. This DOC consumption under the oxidative condition might enhance the precipitation of Fe separated organic ligands. The soil-stream interface suggested in the present study would be the hyporheic zone as shown by Hill et al. (1998) and Triska et al. (1989).

Changes in Al and Si Concentrations through the Water Movement in the Riparian Zone

In the southern watershed, higher Al concentration at the upstream (S-STu) might be caused partially by the pH lower than those at the other streams (S-STm and -STd) (Table 3). As Fe, Al would be dissolved through the chelation by the organic ligands (Trettin et al. 1997). As a step former to the chelation, the oxidation and the subsequent acid generation in the soil might be related also to the Al dissolution in the soil water and groundwater.

Si concentrations were low in the surface soil where DOC concentrations were high (Figure 14). Furthermore, Si concentrations became higher as approaching to the stream channel and to the deeper soil layer where DOC concentrations were low. Si density in a soil layer would thin as organic undecomposed carbon accumulates in its layer, leading to the contrastive regime of Si and DOC dissolution in the soil water and ground water. In the southern watershed, where a riparian wetland had developed, this contrastive regime between Si and DOC dissolution in the water sources and pathways might be reflected in the opposite longitudinal patterns of the concentration changes in the stream water.

Longitudinal Patterns of the Na^+ and Cl^- Concentrations in the Stream Water

In the northern watershed, downstream decreases of Na^+ and Cl^- concentrations were caused by the inflow of the surrounding soil water and ground water having low concentrations of these two ions into the stream channel, suggesting the relatively simple dilution (Figure 16 and 17). In the southern watershed, Na^+ concentration in the stream water would be diluted and decreased downstream, as in the northern watershed (Figure 16).

Contrarily, Cl^- concentrations in the stream water in the southern watershed would increase downstream owing to the inflow of the surrounding soil water and ground water having high Cl^- concentrations into the stream channel (Figure 17). In the southern watershed, moreover, the spring Cl^- concentration in the stream water (S-STu) was three-fold lower than that in the northern watershed (N-STusp) (Table 3); the Cl^- concentration in the southern stream was subject to the concentration increase by the surrounding water more than that in the northern watershed. This difference in Cl^- concentration in the spring water was more prominent than those in the other solutes (Table 3), except for NO_3^- which had very low concentrations in the groundwater. Therefore, the Cl^- concentration in the spring water would be influenced largely by the factors other than the riparian zone.

ACKNOWLEDGEMENTS

We would like to thank all the staffs of the Hokkaido Forest Research Station, FSERC, Kyoto University. The present study was facilitated by the Forest Research Station, Field Science Center for Northern Biosphere, Hokkaido University. T. Kuboi helped me to install groundwater wells. T.

Watanabe supported the collections of the related literatures. Especially, Dr. H. Shibata, Dr. F. Satoh and Dr. K. Sasa, some researchers (Dr. SF Christopher and X. Xu), and some graduate students, Hokkaido Univ. gave helpful and useful suggestions to the present study. The present study was also supported by the JSPS Research Fellowships for Young Scientist.

REFERENCES

Arai T (1994) *Field method in environmental hydrology* (in Japanese). Kokonsyoin, Tokyo.

Bowden WB (1987) The biogeochemistry of nitrogen in freshwater wetlands. *Biogeochemistry* 4: 313-348.

Choi J, Hulseapple SM, Conklin MH, Harvey JW (1998) Modeling CO_2 degassing and pH in a stream-aquifer system. *J Hydrol* 209:297-310.

Dalva M, Moore TR (1991) Sources and sinks of dissolved carbon in a forested swamp catchment. *Biogeochemistry* 15:1-19.

FAO/Unesco (1974) *Soil map of the world.* 1:5,000,000, vol. I, Legend. Unesco.

Ford TE, Naiman RJ (1989) Groundwater-surface water relationships in boreal forest watersheds: dissolved organic carbon and inorganic nutrient dynamics. *Can J Fish Aqua* Sci 46: 41-49.

FSERC, Kyoto University. http://www9.ocn.ne.jp/~sibe/index.html (in Japanese).

Hart SC, Nason GE, Myrold DD, Perry DA (1994) Dynamics of gross nitrogen transformations in an old-growth forest: The carbon connection. *Ecology* 75: 880-891.

Hedin LO, von Fischer JC, Ostrom NE, Kennedy BP, Brown MG, Robertson GP (1998) Thermodynamics constraints on nitrogen transformations and other biogeochemical processes at soil-stream interfaces. *Ecology* 79: 684-703.

Hill AR, Labadia CF, Sanmugadas K (1998) Hyporheic zone hydrology and nitrogen dynamics in relation to the streambed topography of a N-rich stream. *Biogeochemistry* 42: 285-310.

Jones JB Jr., Mulholland PJ (1998) Carbon dioxide variation in a hardwood forest stream: An integrative measure of whole catchment soil respiration. *Ecosystems* 1:183-196.

Kaneko S, Hirano Y, Fukayama T, Furusawa H. (2003) Temporal variation of dissolved iron and aluminum concentrations of stream water in forested catchments in Kyoto Prefecture. Ap. For. Sci. 12: 93-97 (in Japanese with English summary).

Machida H, Arai F (1992) *Atras of Tephra in and around Japan,* 15. University of Tokyo Press. Tokyo, p. 276 (in Japanese).

Martin JH, Fitzwater SE (1988) Iron deficiency limits phytoplankton growth in the north-east Pacific subarctic. *Nature* 331: 341-343.

Martin JH, Fitzwater SE, Gordon RM (1990) Iron deficiency limits phytoplankton growth in Antarctic water. *Globlal Biogeochemical Cycles* 4: 5-12.

Nakagawa Y, Shibata H, Satoh F, Sasa K (2008) Riparian control on NO_3^-, DOC, and dissolved Fe concentrations in mountainous streams, northern Japan. *Limnology* 9: 195-206.

Nakagawa Y, Shibata H, Satoh F, Sasa K (2009) Influence of CO_2 partial pressure and other factors on spatial and temporal variation of pH in two streams draining the watersheds having different size of riparian zone. *Environmental Science* 22: 173-186.

Nelson PN, Baldock JA, Oades JM (1993) Concentration and composition of dissolved organic carbon in streams in relation to catchment soil properties. *Biogeochemistry* 19: 27-50.

Ohte N, Tokuchi N, Suzuki M (1995) Biogeochemical influences on the determination of water chemistry in a temperate forest basin: Factors determinig the pH value. *Water Resour Res* 31:2823-2834.

Shibata H, Sugawara O, Toyoshima H, Wondzell SM, Nakamura F, Kasahara T, Swanson FJ, Sasa K (2004) Nitrogen dynamics in the hyporheic zone of a forested stream during a small storm, Hokkaido, Japan. *Biogeochemistry 69*: 83-104.

Stumm W, Morgan JJ (1981) *Aquatic Chemistry*, 2nd ed. John Wiley and Sons Inc. New York.

Suzuki Y, Kuma K, Kudo I, Hasebe K, Matsunaga K (1992) Existence of stable Fe(II) complex in oxic river water and its determination. *Wat Res* 26: 1421-1424.

Takeuchi M (1980) Studies on the freezing and thawing of the volcanic ash soils in eastern Hokkaido. I. The seasonal aspects of the freezing and thawing in a plowed field. *Bull Kyoto Univ Forests* 52: 117-129. (in Japanese with English summary).

The Japan Society for Analytical Chemistry, Hokkaido Branch (1994) *Mizu no bunseki 4th ed*. Kagaku Dojin. Kyoto (in Japanese).

Trettin CC, Jurgensen MF, McLaughlin JW, Gale MR. (1997) *Effects of forest management on wetland functions in a sub-boreal swamp.* In Trettin CC,

Jurgensen MF, Grigal DF, Gale MR, Jeglum JR (eds.) Northern forested wetlands: Ecology and management. *CRC Press*. 411-428.

Triska FJ, Kennedy VC, Avanzino RJ., Zellweger GW, Bencala KE (1989) Retention and transport of nutrients in a third-order stream in northwestern California: Hyporheic processes. *Ecology* 70: 1893-1905.

In: Riparian Zones
Editor: A. Kerem et al.
ISBN 978-1-61942-677-1

Chapter 5

PROTECTION OF RIPARIAN LANDSCAPES IN ISRAEL

Tseira Maruani and Irit Amit-Cohen
Department of Geography and Environment,
Bar-Ilan University, Ramat-Gan, Isral

ABSTRACT

Riparian landscapes are natural habitats of unique ecological and scenic values, which are highly sensitive to human intervention and impact. Yet, due to their qualities, and especially the presence of water, they are also usually attractive for recreation purposes. This is more so in arid and semi-arid zones like Israel. Nevertheless, in the past, the importance of riparian landscapes in Israel did not receive adequate attention in policy and planning. As a result, over the years they were exposed to various negative impacts, including pollution by industrial and agricultural effluents, exploitation of water for agricultural and other purposes, and land use conflicts. Although in recent years with the growing awareness of their ecological and recreational potential, considerable efforts are being invested in the rehabilitation of deteriorated riparian landscapes, their protection is still deficient.

This chapter reviews and examines policy tools used for the protection of riparian landscapes in Israel, based mainly on regulations, reports and existing literature. It concludes by offering some lessons for policy-making in general and suggestions for improving the protection of riparian landscapes in Israel in particular.

Keywords: environmental policy, river corridors, conservation, open spaces.

INTRODUCTION

Riparian landscapes are natural habitats of unique ecological and scenic values, which are highly sensitive to human intervention and impact (Goren, 2000). Yet, due to their qualities, and especially the presence of water, they are also usually attractive for development as well as for recreation purposes. This is more so in arid and semi-arid zones like Israel (Burmil, 1999; Patten, 1998). Nevertheless, in the past, the values of riparian landscapes in Israel did not receive adequate attention in policy and physical planning. As a result, they were exposed over the years to various negative impacts, including pollution by industrial and agricultural effluents and household sewage, exploitation of water for agricultural and other purposes, political disputes and land use conflicts (Amir-Shapira and Feldman, 1999; Gabbay, 1998). In recent years, mainly since the year 2000, with the growing awareness of their ecological and recreational potential, considerable efforts are being invested in the rehabilitation of deteriorated riparian landscapes. Nonetheless, according to a recent report most of the rivers in Israel and their surroundings still suffer from pollution and other negative impacts (Amir-Shapira, 2007).

This chapter reviews the protection of riparian landscapes in Israel. It presents the policy tools used for protection of riparian landscapes up to the year 2000, examines their effectivity and reveals embedded deficiencies. The methodology is based on the review and content analysis of various sources, mainly laws, reports and literature. The first part of the chapter introduces the importance of riparian landscapes in general, covering among others ecological, environmental and scenic aspects. It than presents the subject of riparian landscapes in Israel and the factors that affected policy and priorities concerning the use of water in general and protection of riparian landscapes in particular. The second part reviews and examines protection tools, focusing on legislation, institutional structure and physical planning. The third and last part of the chapter discusses the deficiencies and weaknesses that were revealed and proposes improvements to the existing state.

THE IMPORTANCE OF RIPARIAN LANDSCAPES

According to Naiman and Décamps (1997) a riparian zone encompasses the stream channel and that adjacent portion of the terrestrial landscape from the high water mark toward the uplands, where vegetation might be influenced by elevated water tables or flooding. The width of a riparian zone and the diversity of its functional attributes are related to the size of the stream, its position within the drainage network, the hydrologic regime and the local geomorphology (Gregory et al., 1991; Naiman and Décamps, 1997). In other words, a riparian zone may be regarded as a linear physical landscape entity composed of aquatic and terrestrial components and the interface between them. Riparian landscapes are unique ecosystems, highly important ecologically and environmentally, and they constitute a visually and functionally outstanding component of the open space system. They form complex habitats, richer than the average with species of plants, animals and microorganisms, thus contributing to overall biodiversity (Gregory et al., 1991; Naiman et al., 1993). These habitats and the richness of species they sustain are especially sensitive to various impacts and interferences, including seasonal or other changes in water physical qualities (like temperature, salinity or electric mobility) or quantities (Allan, 2004; Goren, 2000; Karr, 1994; Naiman and Décamps, 1997; Pollock et al., 1998; Shandas, 2007; Stauffer and Best, 1980). They are considered the most diverse, dynamic and complex habitats (Goren, 2000; Gregory et al., 1991; Naiman and Décamps, 1997).

Riparian ecosystems perform a variety of environmental and ecological services (Allan, 2004; Brauman et al., 2007; Hale and Adams, 2007; Naiman and Décamps, 1997; Patten, 1998). Many of these are considered life-supporting systems (De Groot, 1992; Naveh, 1997), although they are difficult to evaluate in economic terms (Chavas, 2000; Constanza, 2000). Due to their linear nature, riparian landscapes are natural ecological corridors, allowing connectivity between habitats and patches in the landscape (Beier and Noss, 1998; Bentrup and Kellerman, 2004; Jongman, 1995; Machtans et al., 1996; Naiman et al., 1993; Ndubisi et al., 1995; Shkedi and Sadot, 2000; Taylor et al., 1995; Walmsley, 2006; Weber et al., 2006). This is especially important in populated areas, where riparian landscapes may be the last remnants of an open natural space within a built-up area. Another function that is especially important in populated areas is their potential as catchement basins for floods, thus avoiding damage to property and lives (Brody et al., 2007; McHarg, 1969; State Comptroller, 1993). They are also attractive for recreation and leisure activities, and due to their linear form are natural candidates for greenway planning, combining opportunities for recreation with conservation of nature, landscape and heritage values (Ahern, 1995; Baschak and Brown, 1995; Bryant, 2006; Fábos, 2004, 1995; Jim and Chen, 2003; Lewis, 1964; Li et al., 2005; Little, 1990). Open lands along rivers and streams are also often used for agriculture, because of their fertility, due to embedded sediments, and relatively level plains (Bentrup and Kellerman, 2004; Karr, 1994).

Riparian landscapes are extremely vulnerable to human impact. Surface water from all over the drainage basin might carry with it soil particles and various pollutants, including pesticides, fertilizers and agricultural wastes into the stream. Their potential for conservation, recreation or agriculture is diminished by conflicts and liabilities from various sources. Pollution by industrial and agricultural effluents and overflow of sewage from treatment plans harms the landscape, disqualifies water for irrigation, prevents recreational water-related activities (e.g. swimming and boating) and repels potential users in general. Exploiting river water for irrigation decreases the availability of water for natural system functions. Though riparian landscapes are attractive for development and recreation, they may be contradictory to certain types of agricultural uses. Moreover, floods may cause damage to agriculture and other uses within the flooded area.

Nevertheless, the presence of water is a potentially dominant factor that attracts development. Unfortunately, often the resulting development ends up with the construction of buildings and infrastructure too close to the water line, thus interfering with the ecological functions of the riparian system.

Considering their uniqueness and attractiveness on the one hand and their vulnerability to various impacts on the other, riparian landscapes are in need of effective protection from inappropriate uses. In Israel, a hot and dry land, such landscapes are especially important. The next section describes the factors that affect rivers and riparian landscapes in Israel.

THE RIVERS IN ISRAEL

The priorities concerning water and riparian landscapes in Israel stem from physical conditions – mainly geomorphology and climate – combined with ideological and geopolitical factors.

Israel is situated on the eastern side of the Mediterranean basin, with geomorphology and climate changing both along the north-south axis and the east-west one. The north and center of the country lie within the Mediterranean climate zone, characterized by a short rainy winter and a long, hot and dry summer while the south of the country is more desert-like. The mountainous northern areas are the rainiest with an average of 800-1000 millimeters per year, decreasing in the central zone the a yearly average of 500-600 millimeters. Eastward and southward the average decreases to 100 millimeters and less. These quantities may fluctuate from year to year, and, in addition, there is high occurrence of droughts. All these factors add up to a continual condition of water shortage (Menahem, 1999).

The area of the State of Israel is intersected longitudinally by a series of mountain ranges that divide it into an eastern basin, where rivers flow towards the Jordan River, the Sea of Galilee and the Dead Sea, along the natural eastern borders, and a western basin, where rivers flow into the Mediterranean Sea. The western basin's climate is milder and relatively rainy, while the eastern basin's is hot and dry. Due to the limited rain quantities, many river sections are in fact seasonal streams, flowing during winter and spring, and drying out in the summer until the next rainy season. In the dry south, the streams are characterized by short strong flows following rain events, which dry out quickly. Only a few rivers that are fed by year-round springs have water flowing in all seasons.

Seasonal streams are exceptionally sensitive because of the fluctuations in water availability for habitat performance in addition to the fact that some of them do not enjoy a strong flow even in the rainy season. These characteristics affect their public conceptual image as landscapes fit for conservation or suitable for recreation purposes. River sections that run through areas declared as nature reserves or national parks are protected, along with the whole area, under the National Parks and Nature Reserves Law of 1963. However, a river or stream crossing an area that is not characterized by special scenic or nature values were usually not grasped as deserving protection. Consequently, they are under pressures for development, especially in areas of level topography, such as the inner and the coastal plains.

The most important rivers, considering water discharge amounts and landscape impact are in Israel's western basin, crossing the coastal plain where the majority of the population (about 80%) is concentrated. These also have the highest potential for recreation. However, their proximity to densely settled areas has resulted in their deterioration due to various negative impacts, such as: trapping riparian waters for agricultural use and other purposes, pollution by agricultural and industrial effluents, overflow of sewage from treatment facilities, construction or agricultural cultivation close to the water line, neglect and garbage disposal (Amir-Shapira and Feldman, 1999).

In addition to this, because of the scarcity of water resources and frequent droughts, the national water policy focused on keeping control over all water sources and prioritizing the use of water for agriculture and other uses. For example, since the mid 1950s, most of the water from the Yarkon springs was captured and transported to agricultural fields in the Negev (the southern region of the land) through the Yarkon-Negev pipeline; as a result, the Yarkon River, the main river in the core, densely-populated area of Israel - once a wide deep stream that the British soldiers had to cross by boats when conquering the land from the Turks in the year 1917 – deteriorated into a narrow and shallow stream, where most of the flow consisted of industrial effluents and initially-treated sewage. The negative impacts on riparian landscapes call for examination of existing protection and management tools and their effectivity.

Protection of Riparian Landscapes in Israeli Legislation

The protection of landscapes by legislation is a universally accepted model for the conservation of outstanding scenic or natural values (Maruani and Amit-Cohen, 2007). Legislation is a source of authority and constitutes a strongly effective policy tool.

Yet this tool has its drawbacks, among them low flexibility, requiring a long, complicated bureaucratic process whenever a change in legislation is desired. This is also a source of conflict between conservationists and developers and landowners. On the other hand, protection based on a statutory declaration is stronger than any other protection tool, and provides the relevant authorities with enforcement measures.

In Israel, legislation concerning water resources, including rivers, is complex, involving a multitude of laws and regulations as well as a multitude of authorities and organizations that are meant to enforce them (Kaplan, 2004). This section presents three laws that are the most relevant to the protection of riparian landscapes, as follows: the Water Law, the Drainage Law and the Rivers Authorities Law.

The Water Law

The Water Law, enacted in 1959, should be understood in the context of the general scarcity of water resources in Israel, the relative average low rainfall and the frequent droughts, on the one hand, and the centralistic governance style and intensive involvement of the State, on the other (Menahem, 1999). The main objectives of the law were to secure the State's sovereignty over all water sources and ensure their utilization for the benefit of Israeli society. The law determines that "... the water sources in the State are owned by the public, controlled by the State and designated for its population and development needs. The water sources for that matter are the springs, the streams, the rivers..." etc. (Water Law, sec.1-2). The law regulates the management of water sources, including allocation for users and preservation of water quality. For these purposes, the law created several institutions, among them a National Water Board, headed by the Water Commissioner. The law gave priority to the agricultural sector, which at the time was regarded not only as a leading sector in the national economy but also as the ideological and political elite of Israeli society (Menahem, 1999; Schiffman, 1999). This was reflected in various attributes of the law, among them its subordination to the Minister of Agriculture, including the power to appoint the Water Commissioner, who was to lead the design and implementation of water policy in Israel. Thus, it is no wonder that the first seven elected commissioners were from the agricultural sector, as were also most of the National Water Board members (Menahem, 1999). Only forty years later, in the 1990s, following ideological, political and social change processes in Israeli society, including the decrease in the economic and ideological importance of agriculture and the acceleration of development (Maruani, 2005; Schiffman, 1999), the Water Law was transferred to the authority of the Minister of National Infrastructures and some of its powers were split among several other institutional structures. Laster (2004) claims that this split in authorities is the reason for the decrease in water quality, in contrast to Western Europe where a centralistic approach was adopted.

The Water Commissioner, among others, used his powers collect water from springs and rivers, and allocate it to consumers, thus reducing the natural flow, ignoring the ecological consequences, which, in turn, also had a negative impact on the attractiveness and availability of the surrounding area for recreation (Laster, 1995).

The only reference in the Water Law to the protection of land near water sources is the available option to define buffer strips for purposes such as

"...protection of water, of water source ...", that are wide "... no more than is necessary for the purpose of the buffer strip..." (sec.14-15). No references are made to purposes involved in the protection of landscape or other values embedded in the terrestrial area outside the water itself.

The Water Law reflects a utilitarian approach towards water resources, and almost completely ignores other values. Some years ago the law was amended, and to the list of uses for water allocation was added the "...protection and rehabilitation of nature and landscape values, including springs, rivers and wetland habitats" (sec.6). However, there is no other reference to such a goal anywhere else in the law, and its implementation is totally dependent on the awareness and the good will on the part of the decision makers. It should also be noted that the interests of the Head of the Water Authority (who replaced the original Water Commissioner, following an amendment in 2006) and the Ministry of National Infrastructures do not coincide with the interests of landscape conservation. This was even truer in the past, when the main interest of the Water Commissioner and the Ministry of Agriculture was the supply of water to the agricultural sector.

The Drainage Law

The Drainage Law (or in its full name: the Drainage and Flood Protection Law) was legislated in 1957, after floods in the mid 1950s covered vast areas, mainly agricultural fields, causing severe economic damage. The objective of the law was to prevent the recurrence of floods by forming suitable institutions to manage drainage. This law, too, was subordinated to the Minister of Agriculture – and still is - since agricultural lands were conceived as being most threatened by potential floods. The priority given to agriculture is expressed in the first section of the law, which defines "drainage" as "...any operation intended to concentrate, to store, to carry or to remove surface or any other water which harm or may harm agriculture, public health..." etc.

The Drainage Law states that the "...Minister of Agriculture may... establish a drainage authority..." (Drainage Law, sec.11), but the establishment of such an authority is not obligatory. The law also specifies the duties and powers of such an authority, including regulating drainage and initiating drainage projects within its jurisdiction. The roles of a drainage authority as they are specified and defined in the law do not refer to the protection of the relevant riparian landscapes in any other way. In addition to drainage authorities, the law determines the establishment of a National

Drainage Board to advise the Minister on matters of drainage, such as the declaration of drainage zones or the approval of drainage projects (sec.2). This is an obligatory body, complementing the relevant institutional structure for regulating drainage and preventing floods.

The Drainage Law defines some relevant terms for its purposes including "channel" and "buffer strip." A Channel is defined as "...river, stream... and any other water route... where water is flowing or standing always or occasionally." A buffer strip is defined as "strips of land along both sides of a channel (sec.1). The law prohibits agricultural cultivation or construction within a buffer strip. However, according to the law, the overall width of buffer strips on both sides should not exceed half the channel's width, and no more than 5 meters each (sec.5-6). It is needless to say that the potential protection of the riparian habitat is very limited given such conditions.

The Drainage Law, like the Water Law, expresses the priority given to agriculture. Most drainage projects during the 1960s and 1970s were intended to solve drainage problems in agricultural and open areas, although in populated areas potential damage from floods to property and persons is much higher. Moreover, since the legislation of the Drainage Law in 1957, the scope of development for residential, occupational and infrastructure uses increased greatly, much of it was at the expense of open and agricultural lands where rains could have previously penetrated the soil. As a result, the amounts and intensity of surface flow towards rivers increased considerably, and with them the risk of floods. In spite of that, the Drainage Law was not updated to include instructions regarding measures for the control of drainage in new development plans. This neglect resulted in recurrent flood events in various areas, causing millions dollars worth of damage (Laster, 2004; State Comptroller, 1993, 1999b).

The Rivers Authorities Law

The Rivers Authorities Law (RA Law: the full name being "Rivers and Springs Authorities Law") from 1965 complements the Water Law and the Drainage Law by referring not only to the protection of the water but also to the land adjacent to it, thus expressing the values of riparian landscapes for the first time in Israeli legislation. This expression is also reflected in the ministerial subordination; when legislated, the RA Law was subordinated to both the Ministry of Interior and the Ministry of Agriculture; however, it was transferred to the Ministry of Environment after it was created in 1989.

The RA Law states that the "…Minister may… establish an authority for a certain river or part of it… or impose on a drainage authority powers of a river authority…" (RA Law, sec.2). That is to say, a river authority – like a drainage authority – is not obligatory. Among the duties of a river authority, the law counts "protection of the landscape and nature values along the river on both sides…" but also regulation of the river's water flow and drainage within its jurisdiction (sec.3).

This means that there is an overlap between the duties of a Drainage Authority and that of a River Authority, albeit they are related to different ministries. Laster (2004) claims that the law was intended to manage the rivers on a watershed approach basis, but actually it did not enable that, since it demanded subordination to the Water Law and the Water Commissioner. Moreover, the jurisdiction of a River Authority does not necessarily cover a whole watershed. For example, the jurisdiction of the Yarkon River Authority was limited to 20 meters on each side of the river.

The RA Law was implemented for the first time in the year 1988 – 23 years after its legislation – with the establishment of the Yarkon River Authority. It took six more years to establish the Kishon River Authority, in 1994, which was the second, and, up–to-now the latest authority based on this law. Laster (2004) comments that the reason it took so long to implement the law was the original subordination to two ministers. Another criticism of the law relates to the absence of a national body, something like the National Drainage Board. In other words, the RA Law represents a particularistic approach, referring to each river separately and lacking a broad vision on a national scale.

INSTITUTIONAL STRUCTURE

While legislation is a source of regulative powers, there is a need for institutions to use these powers and enforce the regulations. In Israel, there are several institutions that have to do with management and protection of riparian landscapes, the most important of them being the Drainage Authorities, the River Authorities and the Rivers Administration. They differ by their status, composition, main objectives, powers and budgeting.

Drainage Authorities

Drainage Authorities (DAs) may be established by the Minister of Agriculture, as already mentioned, subject to the consent of the Minister of Interior and the relevant local municipalities, which are also represented in its composition. The DA is responsible for managing drainage zones within its jurisdiction, including initiating, developing and maintaining drainage projects. The powers of a DA may contribute greatly to the protection of the relevant riparian landscapes, since managing drainage requires, among other things, restrictions on development along the water route.

Following the legislation of the Drainage Law, the Minister of Agriculture issued an ordinance in 1960 establishing 26 DAs. The ordinance specified the jurisdiction allocated for each DA, generally consisting of low-elevated plains, mainly agricultural lands (State Comptroller, 1993). Laster (1995, 2004) comments that the large number of DAs was due to political pressures rather than hydrological needs. Most of them actually operated as organs of the relevant Regional Councils - which are the local municipalities in the rural zones – and were dominated by representatives of the agricultural sector, thus reinforcing its control over water sources.

The recurrence of severe flood events implies the existence of shortcomings and deficiencies in the Drainage Law's implementation in general and in DAs' operation in particular. Especially memorable are the floods of winter 1991-2, when river overflows flooded large areas all around Israel, disrupting the course of everyday life: people had to be temporarily evacuated from their homes, roads became impassable, large agricultural areas were under water and there were even some fatal casualties. The overall direct and indirect damage to households, businesses, public property and agriculture amounted to tens of millions of dollars. Following those events, the State Comptroller (1993) issued a report that pointed out various faults in the DAs' functioning, stating among others that the drainage infrastructure had been neglected over the years and the sums required for proper planning, regulating and maintenance of many rivers were not allocated and invested as should have been done.

In 1996, after studying the State Comptroller's report, the Minister of Agriculture issued an ordinance reorganizing the DA system. Their number was decreased from 26 to 11, each covering a jurisdiction overlapping a natural drainage basin, and together covering the whole country. Each DA is directed independently of the Regional Council system, intended to apply a whole watershed approach rather than serve local interests (Laster, 2004).

Although since this reorganization floods have reccurred, it seems that - especially since 2000 - DAs are taking more active measures to prevent flooding, revealing a more environmentally-oriented approach, including the conservation of landscape and ecological values in the riparian zones, in collaboration with other institutions and organizations, such as the Ministry of Environment, the Nature and Parks Authority and the Society for the Protection of Nature (Carmel Drainage Authority 2004).

Rivers Authorities

A River Authority (RA), with powers to limit and control development along the river, is a potentially effective tool in the protection of riparian landscapes. However, this potential cannot be fully realized, partly due to constraints imposed by the RA Law, such as the limited size of the area under the RA's jurisdiction. The RA's effectivity is also limited by its obligation to comply with the Water Authority (RA Law, sec.4), even in cases when their interests are contradictory. An RA's composition is more complex than a DA's, including several government officials. Thus, conflicts between government officials and local municipalities' representatives may also hamper the RA's functioning. Another potential source of conflict is the existence of a DA in the same jurisdiction since the duties and responsibilities according to the Drainage Law and the RA Law partly overlap.

The RA Law states that an RA is appointed to manage a particular river and is supposed to operate independently, following its management's objectives and policies. However, there is lack of a master organization on a national scale to design and lead a comprehensive policy towards the protection of riparian zones. With the absence of national policy, the influence of local interests increases, thus allowing development and infrastructure on or adjacent to riverbanks that should have been preserved.

Nevertheless, an RA is still the only institutional structure that is essentially specified for riparian nature and landscape protection in Israel. This may be exemplified by the Yarkon River Authority (YRA), which since its establishment in 1988 has initiated a master plan for the Yarkon River, seeking, among other goals, to restore the riparian ecosystem and contribute to environmental quality, aesthetic values and climate amelioration (Rachmimov 1996). The YRA develops bicycle trails and areas for recreation along the river and is currently monitoring water quality etc. The YRA was, in fact, a pioneer and a model in river restoration in Israel. Regretfully, however, only one more RA was established till now: the Kishon River Authority (in 1994). Both the Yarkon and the Kishon are large rivers, passing through densely populated areas - the Yarkon in the Tel Aviv metropolis, in the center of the State, and the Kishon in the northern Haifa metropolis - that were in a severely deteriorated state due to pollution from multiple sources, water trapping, etc. However, several other large rivers in a similar condition could have profited from a similar specified institutional structure to manage and restore them.

The Rivers Administration

The Rivers Administration (RAD) was established in 1993 as a mutual initiative of the Ministry of Environment and the Jewish National Fund (JNF), the latter being a historical non-governmental organization that is currently intensively involved in forestry and outdoor recreation. Other institutions and organizations that take part in the RAD are the Nature and Parks Authority, the Ministry of Interior, the Ministry of Agriculture, the Water Authority, the Israel Government Tourist Corporation and the Society for the Protection of Nature which is a non-governmental organization (Amir-Shapira and Feldman, 1999; Gabbay 1998). The declared goals of RAD include: conservation of high-quality open spaces along river channels, controlled development of recreation and tourism foci within riparian zones and rehabilitation of rivers to a state enabling sustainability of unique habitats and landscape values (Amir-Shapira and Feldman, 1999). RAD initiates and promotes plans for the restoration of rivers, issues guidelines for existing and future River Authorities, supports the establishment of local river administrations and is actually involved in many restoration projects that are in progress, especially since 2000. For instance, the RAD takes part in the steering committee of the successful restoration project for the Alexander River – a large coastal river in the Sharon region, and one of the first to enter a restoration program along with the Yarkon River. This is done in collaboration with many other bodies, including the Alexander Drainage Authority, the Emek Hefer Regional Council, the Emek Hefer Streams Society, the Water Authority, the JNF, the Israel Government Tourist Corporation, the Nature Conservation Research Institute, the Ministry of Environment, the Hadera Environmental Association, the Nature and Parks Authority, the Society for the Protection of Nature, the District Planning Commission, the Ministry of Interior and the Israel Lands Administration (Cohen-Shoel 2003). It is clear that when so many organizations are involved, conflicts and clashes of interest are unavoidable and may hamper conservation efforts. For example, the Israel Lands Administration is supposed to manage all public lands, which in Israel amount to 93% of the area of the State. However, since the 1990s, it actually acts as the State's main developer, initiating large intensive development plans wherever possible, driven by economic interests (Maruani, 2005). Such a development-oriented organization is bound to confront conservation-oriented bodies like the Nature and Parks Authority and the Society for Protection of Nature. The RAD, which is guided by a statewide vision regarding rivers and riparian landscapes, can bridge such controversies and promote restoration and conservation efforts.

Nevertheless, the protection offered by the RAD is limited since this is not a statutory body, and consequently lacks regulative and enforcement powers. Whenever action against polluters or other offenders of the riparian landscape is needed, it is the Ministry of Environment that is supposed to interfere and take steps towards enforcement. Also, due to lack of obligatory organizational procedures, RAD's operations are spatially and temporally not consistent in essence, scope and timing. Therefore, in many cases, riparian landscapes in Israel are still subject to the arbitrariness of local authorities, and are not consistently regarded as worthy of protection. In addition, the RAD's potential effectiveness is reduced by budget limitations, since budget allocated for riparian management is divided among too many organizations, including DAs and RAs (Laster, 2004).

The lack of a statutory basis is a source of instability for the RAD, since it may be dismantled by a ministerial decision as quickly as it was established. Nonetheless, since its establishment in 1993, the RAD is a dominant factor in river restoration in Israel.

PHYSICAL PLANNING AS A PROTECTION TOOL

Riparian landscapes are integral parts of the land. As such, physical planning of land uses at national, regional and local levels is a potentially strong tool for their protection. In Israel, land uses are determined by a statutory planning system, according to the Planning and Building Law (PBL) of 1965 (which replaced former British mandatory legislation from 1936). The system is three-tiered, with Planning and Building Commissions at national, district and local levels, each having the power to control lower-tier decisions. All planning commissions may initiate outline plans within their jurisdiction.

National Planning

The Sharon Plan: The Sharon Plan was the first comprehensive national master plan prepared after the declaration of statehood in 1948. The plan was initiated and prepared within the (then) Planning Department of the Ministry of Labor and Construction, by a team headed by architect Arie Sharon, and published in 1951. This plan was not statutory; nevertheless it had significant long-range effects on spatial planning in Israel. The plan refers to five main facets of planning: agriculture, industry, transportation, parks and new cities. The agricultural plan is based on a national water policy that wishes to divert water out of relatively rich sources – among them rivers in the north and the Yarkon in the center of the State – and carry them to the dry south, thus enabling agricultural settlements that are regarded in the plan as a key factor in development and economic independence (Sharon, 1951). In other words, the Sharon Plan again reflects the economic and ideological values attached to agriculture, and grasps the water of the rivers as input for agriculture and not as a landscape component to be protected.

The plan for the parks in the Sharon Plan proposed conservation of certain areas of outstanding nature and landscape values. Four out of those were to be established immediately as national parks, among them two mountainous landscapes: Mount Carmel and Mount Jarmak in the Upper Galilee, and two riparian ones: the Falik Park proposed around the Yarkon and Ayalon rivers in the Tel Aviv metropolis, and a park west of Jerusalem based on the Sorek River (Sharon, 1951). However, although the Sharon Plan as well as the actual implementation of the Carmel and the Jarmak parks significantly affected spatial planning for years to come, the parks proposed around the rivers were ignored as was also the call for using river corridors as buffers between built-up areas. The Sorek corridor area, for example, was reduced over the years due to development (Amit-Cohen et al. 2005).

NOP 31: statutory planning in Israel at the national level is delineated in the National Outline Plans (NOPs). Out of almost 40 NOPs prepared up to date, only two represent comprehensive national planning. The first of the two was NOP 31, the Combined National Outline Plan for Construction, Development and Immigration Absorption. NOP 31 was prepared and approved in the early 1990s, when Israel was facing a crisis stemming from a sudden unanticipated increase in population due to mass immigration waves from the former USSR (Alterman, 1995). This was the first comprehensive national statutory plan to incorporate environmental considerations on a large scale, stating objectives like: conservation of nature and landscape resources, preservation of surface water quality, nurturing open spaces – among them riparian landscapes - for recreation, and balancing between development and conservation (Feitelson, 1993). The plan enhanced intensification of metropolitan regions in order to preserve open spaces in the rural zone. Despite these good intentions, it has been claimed that they triggered heavy development pressures in the metropolitan core of Israel, even in areas designated for conservation (Maruani, 2005).

NOP 35: the Combined National Outline Plan for Construction, Development and Conservation NOP 35 was approved in 2005, substituting NOP 31, which became obsolete in 1998. This plan, too, aspires to balance between development and conservation, taking this a step further by dividing the whole country into zones defined as development-oriented or conservation-oriented in varying degrees. Theplan specifically refers to river strips – including the water route and the banks 100 meters on each side – and requires every statutory plan to issue instructions concerning conservation of the river and its habitats, protection of its drainage functions, bank stabilization and free access to the public. This seems promising indeed, but it will take a number of years before its achievements can be assessed. However, past experience teaches us that NOP's instructions are not always kept. One example is the National Outline Plan for the Mediterranean coast, NOP 13, which prohibited construction within 100 meters of the water line. That instruction did not always stand up development pressures (State Comptroller, 1999a).

Sectorial NOPs: most national outline plans are sectorial, each dedicated to a specific subject (e.g. roads, power plants, landfills, etc.). None of the plans that have been prepared until now is designated for the protection of rivers and riparian landscapes, although in some cases, there may be a reference to rivers where this is relevant to the main plan's design. Such is the case of the National Outline Plan for Nature Reserves and National Parks, NOP 8, which offers protection to sections of riparian landscapes that are found within areas designated and declared as a nature reserve or a national park. Yet, the plan does not ensure protection along the whole water route. The National Outline Plan for Forests and Afforestation proposes plantings along riverbanks (Kaplan, 1993), but only in small limited areas, most of them outside of metropolitan zones where demand for recreation is especially high. The National Outline Plan for Tourism, NOP 12, offers another example. It designates some rivers as recreational spaces, among them the Sorek River (Amit-Cohen et al. 2005). Yet, development attractive for tourism is usually intensive and could harm existing riparian ecological and environmental values.

Regional Planning

District Outline Plans (DOPs), like NOPs, are essentially guiding plans, relating to the district under the jurisdiction of a relevant District Commission or part of it. Israel is divided into six administrational districts. Almost all of them had valid DOPs that by the early 1990s were mostly out-of-date, and none of them conceived riparian landscapes as objectives for conservation. An example for the disregard of riparian functions is the Ayalon Highway, crossing Tel Aviv through the route that once was the Ayalon River, leaving a rather narrow channel for winter water flow. Planning of this highway started in the early 1970s and it was opened for use at the beginning of the 1990s. In the winter of 1991/2, parts of it flooded, more than 140 cars were trapped in the flood, and transportation along the highway had to be stopped for some time (State Comptroller, 1993). This has since recurred several times, also causing some casualties.

The Planning Administration in the Ministry of Interior promoted preparation of new DOPs, some of which have already been approved since 2000. The new plans are much more environmentally oriented than the old ones, including special references and instructions for protection and conservation of rivers and riparian landscapes. For example, the new DOP 3/21 for the Central Area District designates "river and its surroundings" for conservation. Still, there are considerable variations between DOPs in their definitions and conservation instructions; some are guiding while others are obligatory, etc. (Maruani, 2005). In other words, the protection offered by DOPs is not consistent in scope and intensity. Moreover, since most current DOPs are relatively new, it will be some more time before their effectiveness in riparian protection can be evaluated.

Local Planning

While national and district outline plans offer general guidelines, Local Outline Plans (LOPs) are more detailed and serve as platform for issuing building permits. The permits are approved and issued by Local Commissions, which are in fact organs of the relevant local municipality, composed of the elected political representatives who constitute the municipal board. Thus, Local Commissions are interested in the promotion of local economic development, even when it is contradictory to conservation needs, and they tend to ignore a broader regional vision. District Commission's control is supposed to minimize the influence of local economic and political interests on land use decisions. However, since riparian landscapes are attractive for housing, it is no wonder that initiatives for development within riparian areas recur, assisted by interested Local Commissions which regardless of their negative impact and potential flood risk (see for instance: Shechori, 1999; Shmul, 2000; Vaserman-Amir, 1999). Development decisions are a major causal factor where floods are concerned (Brody et al., 2007; State Comptroller, 1993). The recurrent flood events in Israel, and their negative impacts and costs indicate that local development interests were not restricted by the District Commissions.

Discussion

The examination of legislative, organizational and planning tools used for protection of riparian landscapes in Israel revealed various faults and deficiencies that may explain the still distressing state of these landscapes. Despite restoration efforts undertaken, mainly since 2000, almost all rivers are still polluted and ecologically deteriorated with only limited sections available and accessible for recreational uses.

Table 1 that presents a comparative assessment of protection tools shows that no one tool can be highly rated on all parameters. In addition, no parameter shows a consistently high rating for all tools. One fundamental problem is that no tool – with the possible exception of RAD - is dedicated to the objective of protecting ecological, environmental and scenic values embedded in riparian landscapes.

For example, each of the laws examined makes a partial contribution to this objective although they differ in aims and scope. However, none of them regards the protection of riparian landscapes as its main aim. In addition, none of these laws reflects a watershed approach, theoretically or practically, albeit activities taking place anywhere in the watershed area, especially development and various pollution generators, eventually affect the river and the landscape along it. Moreover, partial overlap between the aims and directives of the Drainage law and those of RA Law are a constant source of ambiguity and fuzziness as to duties and responsibilities, on one hand, and interorganizational and inter-personal frictions and conflicts, on the other. This situation stands in the way of the conservation interest.

Table 1. Comparison of protection tools

Protection tool		Designation	Relevance	Power	Scale	Implementation
Legislation	Water Law	Low	Medium	High	National	Low
	Drainage Law	High	High	High	National	Low to medium
	River Authorities Law	High	High	High	National	Low to medium
Institutional structure	Drainage Authorities	Medium	High	Medium	Regional	Low
	River Authorities	High	High	Low to medium	Regional	medium Low to
	Rivers Administration	Very high	Very high	Low	National	Medium
Physical planning	Sharon plan	Low	Low	Low	National	Low
	31NOP	Low	Low	Medium to high	National	Low
	35NOP	Low	Low	Medium to high	National	?
	Sectorial NOPs	Low	Low to medium	Medium to high	National	Low
	DOPs	Low	Low	Medium to high	Regional	Low
	LOPs	Varies with plan	Varies with plan	High	Local	Varies with plan

The different tools are interconnected. Legislation determines the structure, responsibilities and operational procedures of the organization intended to implement it. Subjects not covered by the law are left to the discretion of organizational decision makers. Hence, awareness of riparian values on the part of decision makers is an important factor in effective protection. Statutory physical planning, too, relies on legislative power, and therefore inter-relations between laws may affect organizational actions. For example, the Drainage Law does not refer to statutory physical planning, while on the other hand the Planning and Building Law does not require a preliminary examination of possible impacts on drainage before the approval of a new development plan.

The State Comptroller (1993) pointed out that the planning system allowed development too close to water routes and approved plans on large areas without ensuring suitable measures for infiltration of rains within their boundaries. He argues that the severe floods that were the cause for his report could have been prevented, had suitable instructions been embedded in the Planning and Building Law, thus preventing construction within flood retention areas and conditioning plan approval with proper infiltration and drainage solutions.

The protection of riparian landscapes as was formulated and implemented in Israel expresses a particularistic approach, where each river is conceived as a discrete entity, with a separate DA or RA. For comparison, nature and landscape values within areas that have been declared as nature reserves or national parks all over the state are protected centrally and managed by the Nature and Parks Authority, which is a national statutory institution, established by the National Parks and Nature Reserves Law. It seems that centralistic approach was efficient even when development pressures in Israel increased considerably towards the end of the 20th century. The establishment of the RAD indicates a conceptual change towards rivers as well but its lack of statutory position and the multitude of other relevant organizations reduce its effectivity.

The protection of riparian landscapes also reflects the evolution of environmental awareness in Israel (see also: Fletcher, 2000; Vogel, 1999). For example, the Water Law from 1959 almost completely ignores the environmental functions of water sources, including rivers, while the RA Law from 1965 already regards the protection of riparian environmental values as one of its main aims. The legislation also reflects the prevailing priorities in Israeli society at the time, affected greatly by ideological and national security considerations. In the past, agriculture and development were given priority; agriculture was conceived as a main economic basis and development of spatially dispersed small agricultural settlements also grasped as a tool for dominating national space while environmental needs were overlooked (Hershcovich, 2006; Schiffman 1999). It should be noted that all three laws examined were enacted before the global environmental revolution of the late 1960s. Moreover, as a consequence of the Stockholm Convention in 1972 Israel was one of the first nations to establish an environmentally designated institution in the form of the Environmental Service that was established in 1973 within the Prime Minister's office, and later on as part of the Ministry of

Interior. However, the assimilation of environmental awareness was slow, especially among decision makers, until the 1990s when development pressures increased considerably, following a sudden increase of population due to large immigration waves, threatening nature and landscape values, especially in the coastal area and in vicinity of water bodies. This also was one of the triggers for initiation of RAD.

It should also be noted that split authoritative powers, multiplicities and deficiencies that characterize the protection of riparian landscapes in Israel, tend to be typically characteristic of environmental issues, which are naturally complex, interdisciplinary and bound up with economic and social conflicts. In addition, since its independence in 1948, Israel has been facing enormous challenges more than any other developed state, including an unstable geopolitical situation and recurrent wars, absorption of mass immigration waves and serious social conflicts (Alterman, 1995; Fletcher, 2000; Vogel, 1999). Nevertheless, all this still does not explain why former protection frameworks have not been reconstructed in spite of the advances in environmental management and administration in general. For instance, the Drainage Law that was mainly intended to prevent floods is still under the authority of the Ministry of Agriculture although severe damage caused by floods in recent years was mainly in urban areas and not in agricultural fields (State Comptroller, 1993). There is no doubt that the present state of affairs calls for improvement.

CONCLUSION

Riparian landscapes constitute extremely vulnerable ecosystems. They need protection to preserve the unique aquatic habitats with their biodiversity richness and ecological processes as well as their value for scenic and recreational purposes. In spite of that, their protection in Israel is defective, lacking comprehensive suitable legislation, a statutory institutional structure on a national scale and a sectorially designated NOP. The existing state of affairs is characterized by a complex array of authoritative powers, some split and others overlap, and by deficiencies in formulation and implementation of policies. Only limited segments of specific riparian landscapes may, in fact, be regarded as functional healthy ecosystems.

Several lessons can be drawn from the above discussion. We wish to focus here on those that seem the most important and practical for the State of Israel in the immediate future. First, there is need for a revision of present legislation, integrating together existing laws – mainly Drainage Law and RA Law – rephrasing their aims and directives, and rearranging the institutional structure and its powers according to updated environmental and other needs. This should be done considering a whole watershed approach as has already been suggested by Laster (1995).

Second, the prevailing, rather particularistic, approach should be replaced by a comprehensive centralized one, based on a vision and needs on a national scale. This ought to be reflected in all types of protection tools but especially in the institutional structure, as by establishing a national statutory designated organization (similar to the Nature and Parks Authority) or, alternatively, empowering the existing RAD by adequate legislation, including broader definition of its powers and duties.

Third, increasing environmental awareness in general, and awareness of riparian landscapes' ecological, environmental and scenic aspects in particular, are a key element in promoting protection and conservation of such landscapes. This is true for the general public, as has already been claimed by Lowry (1998), but is even more important where decision makers are concerned. Therefore, educational activities, formal and informal, carry great significance for improving and intensifying riparian protection efforts.

Finally, there is hope for riparian landscapes, even when severely deteriorated, providing suitable protection, restoration tools and a plan of action (see for instance: Harnik 2007, Kirk 2005, Tjallingii 2000). This is especially important in a small, densely populated and dry land like Israel.

REFERENCES

Ahern, J. (1995). Greenways as a Planning Strategy. *Landscape and Urban Planning*, 33, 131-155.

Allan, J.D. (2004). Landscapes and riverscapes: the influence of land use on stream ecosystems. *Annual review of Ecology, Evolution and Systematics,* 35, 257-284.

Alterman, R. (1995). Can planning help in time of crisis? Planners' responses to Israel's recent wave of mass immigration. *Journal of the American Planning Association,* 61 (2), 156-177.

Amir-Shapira, D. (2007). Ranking of polluted rivers. Ministry of the Environment,*www.sviva.gov.il/Environment/Static/Binaries/News/rivers_ 2006-7-tavla_1.pdf.*

Amir-Shapira, D. and Feldman, D. (1999). Fresh blood into river veins. *Yarok-Cahol-Lavan*, 26, 4-7 (In Hebrew).

Amit-Cohen I., Cohen, Y., and Cohen, A. (2008). From historical processes to current formation of ecological corridors: Stability and instability of Green Corridors at the margins of Israel's metropolis. Applied Geography (forthcoming).

Baschak, L. A. and Brown, R. D. (1995). An ecological framework for the planning, design and management of urban river greenways. *Landscape and Urban Planning*, 23 (3-4), 211-225.

Beier, P. and Noss, R. F. (1998). Do habitat corridors provide connectivity? *Conservation Biology*, 12, 1241-1252.

Bentrup, G. and Kellerman, T. (2004). Where should buffers go? Modeling riparian habitat connectivity in northeast Kansas. *Journal of Soil and Water Conservation*, 59 (5) 209-215.

Brauman, K. A., Daily, G. C., Duarte, T. K. and Mooney, H. A. (2007). The nature and value of ecosystem services: an overview highlighting hydrologic services. *Annual Review of Environment and Resources*, 32, 67-98.

Brody, S. D., Zahran, S., Maghelal, P., Grover, H. and Highfield W. E. (2007). The rising costs of floods: examining the impact of planning and development decisions on property damage in Florida. *Journal of the American Planning Association*, 73 (3), 330-345.

Bryant, M. M. (2006). Urban landscape conservation and the role of ecological greenways at local and metropolitan scales. *Landscape and Urban Planning,* 76 (1-4), 23-44.

Burmil, S., Daniel, T. C. and Hetherington, J. D. (1999). Human values and perceptions of water in arid landscapes. *Landscape and Urban Planning*, 44, 99-109.

Carmel Drainage Authority, (2004). Drainage and Environment: regularizing rivers with relation to nature, heritage and landscape values. In collaboration with the Ministry of Environment (In Hebrew).

Chavas, J.P. (2000). Ecosystem valuation under uncertainty and irreversibility. *Ecosystems* 3 (1), 11-15.

Cohen-Shoel, N. (2003). The Alexander River. The Hadera environmental Association (In Hebrew).

Constanza, R. (2000). Social goals and the valuation of ecosystem services. Ecosystems 3 (1), 4-10.

De Groot, R.S. (1992). Functions of Nature: Evaluation of Nature in Environmental Planning Management and Decision-Making. Wolters-Noordhoff, Amsterdam.

Fábos J.G. (2004). Greenway planning in the United States: its origins and recent case studies. *Landscape and Urban Planning*, 68 (2-3), 321-342.

Fábos, J. G. (1995). Introduction and overview: the greenway movement, uses and potentials of greenways. *Landscape and Urban Planning*, 33, 1-13.

Feitelson, E. (1993). Open space conservation and immigration absorption in NOP 31. *The Biosphere*, 22 (1), 1-5 (In Hebrew).

Feitelson, E. (1995). Open space conservation in Israel at a turning point. *Horizons in Geography*, 42-43, 7-22 (In Hebrew).

Fletcher, S. (2000). The evolution of coastal management policy in the state of Israel. *Marine Policy*, 24, 395-405.

Frumkin, R. and Kleinhouse, S. (1997). Vegetation changes in Taninim River and the Cabara swamp in the 21st century. *Ecology and Environment,* 4 (2), 65-72 (In Hebrew).

Gabbay, S. (1998). The Environment in Israel. Jerusalem: Ministry of the Environment (In Hebrew).

Goren, M. (2000). Human impact on riparian biodiversity and ways for its preservation and restoration. *Agamit,* 143, 17-21 (In Hebrew).

Gregory, S. V., Swanson, F. J., McKee, W. A. and Cummins, K. W. (1991). An ecosystem perspective of riparian zones. *BioScience*, 41 (8), 540-551.

Hale, B. W. and Adams, M. S. (2007). Ecosystem management and the conservation of river-floodplain systems. *Landscape and Urban Planning*, 80, 23-33.

Harnik, P. (2007). Nine-Mile RE Run. *www.tpl.org/content_documents/Article_Nine_Mile_ReRun.pdf.*

Hershcovich, A. (2006). From the "Sharon Plan" to NOP 35 – the ideology of three national outline plans. *Horizons in Geography*, 67, 31-40 (In Hebrew).

Jim, C. Y. and Chen, S. S. (2003). Comprehensive greenspace planning based on landscape ecology principles in compact Nanjing city, China. *Landscape and Urban Planning*, 65 (3), 95-116.

Jongman, R. H.G. (2001). Ecological networks and greenways in Europe: The twentieth century and beyond. Development of European Landscapes. *Conference proceedings*, Tartu 2001,1, 18-24.

Jongman, R. H. G. (1995). Nature conservation planning in Europe: developing ecological networks. *Landscape and Urban Planning*, 32, 169-183.

Kaplan, M. (2004). The Rivers of Israel: Policy and Planning Principles. Jerusalem: Ministry of the Environment (In Hebrew).

Kaplan, M. (1993). The forest in the 2000s: a national outline plan for forests and forestation- NOP 22. *Ecology and Environment*, 1 (1), 25-31 (In Hebrew).

Karr, J. R. (1994). Landscapes and management for ecological integrity. In Kim, K. C. and Weaver, R. D. (eds.), *Biodiversity and Landscapes* (pp. 229-251). Cambridge university press.

Kirk, D. (2005). Seoul peels back concrete to let a river run freely once again. *www.csmonitor.com/2005/1013/p07s01-woap.html.*

Laster, R. (2004). Law, organization and administration. In Kaplan, M. The Rivers of Israel: Policy and Planning Principles. Jerusalem: Ministry of the Environment (In Hebrew).

Laster, R. (1995). The Yarkon River Authority: the legal framework for its establishment. In Pergament, D. (Ed.) *The Yarkon* (pp. 33-37). Yarkon River Authority (In Hebrew).

Lewis, P. (1964). Quality corridors for Wisconsin. *Landscape Architecture* 54 (2), 100-107.

Li, F., Wang, R. Paulussen, J. and Liu, X. (2005). Comprehensive concept planning of urban greening based on ecological principles: a case study in Beijing, China. *Landscape and Urban Planning*, 72 (4), 325-336.

Little C. E. (1990). Greenways for America. Baltimore: The Johns Hopkins University Press.

Lowry, W. R. (1998). Preserving Public Lands for the Future: The Politics of Intergenerational Goods. Washington D.C.: Georgetown University Press.

Machtans, C. S., Villard, M.-A. and Hannon, S. J. (1996). Use of riparian buffer strip as movement corridors by forest birds. *Conservation Biology*, 10 (5), 1366-1379.

Maruani, T. (2005). Relations between development entrepreneurship characteristics and open space conservation in metropolitan regions. Ph.D. thesis, Bar-Ilan University (In Hebrew).

Maruani, T. and Amit-Cohen, I. (2007). Open space planning models: a review of approaches and methods. *Landscape and Urban Planning*, 81, 1-13.

McHarg I.L.1969. Design with Nature. New York: John Wiley and Sons, Inc.

Menahem, G. (1999). Water policy in Israel: policy paradigms, policy networks and public policy (pp. 35-68). In Nachmias, D. and Menahem, G. (eds.). Public policy in Israel. Jerusalem: The Israel Democracy Institute. (In Hebrew).

Naiman, R. J. and Décamps , H. (1997). The ecology of interfaces: riparian zones. *Annual Review of Ecology Evolution and Systematics*, 28, 621-658.

Naiman, R. J., Décamps , H. and Pollock, M. (1993). The role of riparian corridors in maintaining regional biodiversity. *Ecological Applications*, 3 (2), 209-212.

Naveh Z. (1997). The value of open landscape as life supporting systems. *Israel Environment Bulletin*, 20 (4), 21-24.

Ndubisi, F., DeMeo, T. and Ditto, N.D. (1995). Environmentally sensitive areas: a template for developing greenway corridors. *Landscape and Urban Planning* 33, 159-177.

Patten, D. T. (1998). Riparian ecosystems of semi-arid North America: diversity and human impacts. *Wetlands*, 18 (4), 498-512.

Pollock, M., Naiman, R. J. and Hanley, T. A. (1998). Plant species richness in riparian wetlands – a test of biodiversity theory. *Ecology*, 79 (1), 94-105.

Rachmimov, A. (1996). A Master Plan for the Yarkon River. The Yarkon River Authority (In Hebrew).

Rinat, Z. (1998). Where does the river flow? Ha'aretz 13.1.98 (In Hebrew).

Schiffman, I. (1999). The Limits of Ideology: The Fight for Open Space in Israel (pp. 559-577). In Soden D.L. and Steel, B.S. (eds.), Handbook of Global Environmental Policy and Administration. New York: Marcel Decker.

Shandas, V. (2007). An empirical study of streamside landowners' interest in riparian conservation. *Journal of the American Planning Association*, 73 (2), 173-184.

Sharon, A. (1951). Physical Planning in Israel. Jerusalem: Government Printing office. (In Hebrew).

Shechori, D. (1999). The National Board will limit development along the Yarkon. Ha'aretz 11.4.99 (In Hebrew).

Shkedi, Y. and Sadot, E. (2000). Ecological corridors – a practical conservation tool. Jerusalem: Science Section of the Israel Nature and Parks Authority. (In Hebrew).

Shmul, A. (2000). A plan for 780 housing units on the Kishon River banks towards deposition by the District Commission. Ha'aretz 13.8.00 (In Hebrew).

SPNI, (2001). The court accepted SPNI's appeal and canceled a permit for high-tech building amidst open landscape. Bama, 23 (In Hebrew).

State Comptroller, (1999a). Planning land uses on the Mediterranean coast. Annual report 49, 361-371 (In Hebrew).

State Comptroller, (1999b). The state drainage infrastructure. *Annual report* 49, 242-248 (In Hebrew).

State Comptroller, (1993). Reports on the state drainage infrastructures and the Ayalon Highways planning, management and maintenance of the Ayalon Channel. Jerusalem (In Hebrew).

Stauffer, D. F. and Best, L. B. (1980). Habitat selection by birds of riparian communities: evaluating effects of habitat alterations. *The Journal of Wildlife Management*, 44 (1), 1-15.

Tal, D. (2003a). Drainage forecast for 2003: rivers clogged; roads, fields and houses inundated. Globes 20.1.03 (In Hebrew).

Tal, D. (2003b). Floods in Tel-Aviv and the Dan region: enormous traffic jams, sewers have collapsed. Globes 22.1.03 (In Hebrew).

Taylor J., C. Paine and FitzGibbon, J. (1995). From greenbelts to greenways: four Canadian case studies. *Landscape and Urban Planning*, 33, 47-64.

Tjallingii, S.P. (2000). Ecology on the edge: landscape and ecology between town and country. *Landscape and Urban Planning*, 48 (3-4), 103-119.

Vaserman-Amir, N. (1999). Ministry of Interior and SPNI: too vast building rights in the Yarkon River plan. Globes 9.4.99 (In Hebrew).

Vogel, D. (1999), Israeli environmental policy in comparative perspective. *Israel Affairs*, 5 (2), 246-264.

Walmsley, A. (2006). Greenways: multiplying and diversifying in the 21st century. *Landscape and Urban Planning,* 76 (1-4), 252-290.

Weber, T., Sloan, A. and Wolf, J. (2006). Maryland's green infrastructure assessment: development of a comprehensive approach to land conservation. *Landscape and Urban Planning*, 77, 94-110.

In: Riparian Zones
Editor: A. Kerem et al.
ISBN 978-1-61942-677-1

Chapter 6

Importance of Arbuscular Mycorrhizal Fungi for Recovery of Riparian Sites in Southern Brazil

***Sidney Luiz Stürmer*[1], *Andressa Franzoi Sgrott*[2], *Felipe Luiz Braghirolli*[2], *Rosete Pescador*[1], *Alexandre Uhlmann*[3]**

[1]Universidade Regional de Blumenau (FURB), Departamento de Ciências Naturais. Brazil
[2]Graduation Program in Environmental Engineering, Universidade Regional de Blumenau. Brazil
[3]EMBRAPA Florestas, Estrada da Ribeira, , Brazil

Abstract

The endomycorrhizal association established between arbuscular mycorrhizal fungi (Glomeromycetes) and plant roots are known to have a pivotal role on plant survival and nutrition. In this chapter, we convey that the triad riparian forest-carbon sequestration-arbuscular mycorrhizal fungi should be considered as environmental friendly strategy to recovery riparian sites. Greenhouse and field experiments carried out in soils occurring at the Itajai valley basin have demonstrated the importance of this association for growth and carbon allocation on plants species used to revegetation of riparian sites. Inoculation of pioneer species with selected fungal isolates have shown that more carbon is allocated to roots when compared to non-inoculated plants or plants inoculated with indigenous fungi. For Schinus terebinthifolius we have evidence at field conditions that inoculation with Scutellospora heterogama, Glomus clarum, and Acaulospora koskei increased significantly the quantity of carbon stored in individual plants, calculated using allometric equations. Riparian sites are habitats that harbors a great diversity of glomeromycetes and 42 species has been registered in works developed in the USA and Brazil, most of them in the genera Glomus and Acaulospora. It is also noticeable that all members of Paraglomus (P. occultum, P. brasilianum, P. laccatum) and Archaeospora trappei are commonly found in riparian sites. Considering the role of arbuscular mycorrhizal fungi on growth and carbon accumulation of pioneer plants and the high diversity of glomeromycetes in riparian soils, the inoculation of plants to be used in revegetation process of riparian sites should be viewed as a environmental friendly strategy to recovery these areas and improve carbon sequestration.

INTRODUCTION

Since the beginning of the Industrial Revolution, the composition of planet Earth atmosphere has been changed due to increases on the concentration of greenhouse gases such as carbon dioxide (CO_2) and metane (CH_4) and the impact of the elevation of carbon concentration has been of concern by the scientific community (Desjardins et al. 2005). To sequestrate terrestrial carbon (C) in living plant biomass is one strategy to decrease carbon concentration in the atmosphere and provide a pathway to increase carbon storage in soils (Rees et al. 2005). Increasing of plant biomass is directly dependent on the process of photosynthesis that converts atmosphere CO_2 into organic carbon, and also is influenced by internal plant factors such as leaf age, leaf structure and position, and chlorophyll content (Marabesi 2007). In this scenario is paramount to consider the role of soil organisms establishing mutualist symbiosis with plant species which influence above and belowground plant growth rates (Stulen and der Hertog 1993, Jakobsen and Rosendahl 1990).

Forest ecosystems are considered one of the most precious natural resources that can be used by humankind to provide several goods (*e.g.*, timber, food, wood, etc.) and environmental services, and they are important at the global level in the terrestrial carbon cycle (Caldeira et al. 2002). Deforestation contribute to the greenhouse effect by either releasing C in the atmosphere due to burning and decreasing living plant biomass that function as a sink of atmosphere C (Marcene et al. 2006). In this context, recovery of areas occupied by riparian forest is recognized as important and necessary to preserve one of the most important natural resources for human society: water (Rodrigues and Leitão Filho 2000). WWF (2003) points riparian forests as strategic partners to maintain and to regulate water supply to urban centers and recovery of riparian forests start with revegetation using woody tree species considering their physioecological characteristics, their relations with soil classes and their association with soil mutualists.

The Itajai river watershed is considered one of the most important basins in Brazil. It is located in the state of Santa Catarina, southern Brazil, within an economically very wealthy region. Despite the relatively recent human colonization in this region by Germans and Italians immigrants, riparian forests along the Itajai river are found in advanced stage of environmental degradation. Since 2001, several research activities have been developed to provide the basic knowledge to recover riparian Forest within the Itajai river watershed. In this chapter, we summarize of results of this research agenda concerning arbuscular mycorrhizal fungi (AMF) and the symbiosis they establish with plant species used to recover riparian forests. We first comment on the importance to recover these forests. We also provide information on species richness occurring in riparian soils and include results of plant growth experiments carried out under controlled and field conditions. Throughout this chapter, we convey the idea that the triad riparian forest-carbon sequestration-arbuscular mycorrhizal fungi should be considered as environmental friendly strategy to recovery riparian forests.

Riparian Forests and Soil Carbon Sequestration

Riparian forests are considered vegetation types associated with stream dynamics found along river banks, lakes and watersprings (Ab'Saber 2000) which experience the influence of geology, geomorphology, soils and general climate of the watershed basin (Naiman and Decamps 1997, Ab'Saber 2000, Lima and Zakia 2000). Riparian forests are inundated for part or most of the growing season (Kozlowski 2002) and therefore, the establishment of plant species depends upon their morphological, physiological and reproductive strategies and their interaction with soil organisms to face stresses imposed by the soil flooding and the process of river dynamic (Rodrigues 1992, Naiman and Decamps 1997, Lobo and Joly 2000).

Riparian forests are important not only because of plant biomass production and hence, forest products, but also because they provide a wide range of ecological services (Kozlowski 2002). These forests influence soil hydrology by improving infiltration and percolation processes, decreasing surface runoff and dissipating energy from rain drops upon the soil surface preventing erosion (Bigarella 2003). The typology of riparian forests allows the formation of extensive and continuous areas of vegetation that creates favorable conditions to survival and genetic flow among populations of animals and plants, functioning therefore as ecological corridors (Macedo 1993, Durigan 1994). They also provide shelter for birds and mammals species due to low fire incidence, proximity with agricultural fields and abundant water supply (Kozlowski 2002). From an anthropological view, riparian forests are important to improve water quality, lighten the effects of air and water pollution (Kozlowski 2002). Because of their fertile alluvial soils, their areas are exploited for agriculture and pasture for cattle ranching which causes loss of original vegetation of riparian forests.

The regeneration process of riparian forest brings up a new environmental service: the potential for carbon sequestration. Recovery of areas occupied by riparian forests starts by transplanting seedlings of woody tree species, a process that must consider particularities of each plant species and their relation to edaphic factors. To maximize the input of C in soils it is necessary to maximize plant productivity (Rees et al. 2005) as primary productivity aboveground is the major source of organic matter to most soils (van de Geijn and van Veen 1993). Carbon enters the pool of soil organic matter (SOM) through litter production, incorporation of crop plants after harvesting and via root system (van de Gejin and van Veen 1993). Therefore riparian forest regeneration processes using fast growth species have the potential to promote not only the recovery of the area by revegetation but also to function as atmospheric carbon sink in the plant biomass and soil.

The role of riparian forest revegetation to act as carbon sinks is better understood considering that the carbon allocated to roots is also utilized to maintain symbiosis established by plant roots with soil microorganisms, particularly nitrogen fixing bacteria and arbuscular mycorrhizal fungi (AMF). The latter are keystone component of terrestrial ecosystems forming the arbuscular mycorrhizal association with plant roots. They have been viewed not only as plant symbionts but also as a critical link in the plan-soil continuum (Wilson et al. 2009) as they improve plant nutrition and nutrient uptake (specially phosphorus) and might be an important component of soil organic carbon in addition to facilitate carbon sequestration through stabilizing soil aggregates (Zhu and Miller 2003). In terms of global carbon cycle, it has been estimated that plants can allocate 10-20% of their photosynthate to AMF growth (Jakobsen and Rosendahl 1990) which is a significant amount considering that AMF establishes association with 80% of plant species in nearly every terrestrial ecosystem in the world (Smith and Read 1997, Wang and Qiu 2006). Mycorrhizal fungi can enhance carbon sequestration by distinct mechanisms. AMF external hyphae extending into the bulk soils are expected to translocate C away from the respiratory activity around the roots (Treseder and Allen 2000). Growth and turnover of extra-radical hyphae (ERH) within the bulk soil can influence directly soil carbon dynamics as the residence time of ERH is usually short (Zhu and Miller 2003). ERH also plays an important role in the formation and maintenance of soil aggregates (Wilson et al. 2009); soil aggregation is an important mechanism to protect C-rich compounds from microbial degradation and therefore to improve C sequestration. Glomalin, a recalcitrant protein produced solely by AMF, can be responsible for the production of 4-5% of carbon and nitrogen in some soils (Rillig et al. 2001). Besides these processes, the amount of chitin in the mycelia cell walls represent C allocated to mycorrhizal tissues that could be long-lived in soils (Treseder and Allen 2000). On the other hand, the contribution of asexual spores produced by AMF as carbon sink has not been readily measured.

The role of mycorrhizal fungi in contributing to increase carbon sequestration is evidenced when the successional model for recovery of riparian forests is applied. In this model, pioneer species are pivotal as they will provide conditions for development of non-pioneer species (Kageyama et al. 1990). Pioneer species are characterized by fast growth, shade intolerant and small seeded species (Barbosa 2000). Evidences have been built in the literature that pioneer species are highly responsive to inoculation of AMF. Siqueira et al. (1998) studied the mycotrophic growth and response to superphosphate of 28 woody species belonging to different successional stages and observed that pioneers were markedly more responsive to mycorrhizal and P treatments that climax one. Considering shoot dry biomass, mycorrhizal response of pioneers and climax species over controls were 813% and 0%, respectively. Zangaro et al. (2000) calculated that the mycorrhizal dependency of pioneer species was 90% compared to 48%, 12% and 14% for early secondary, late secondary and climax species, all respectively. They also showed that greenhouse and field mycorrhizal colonization tended to be higher in pioneer species compared to climax species. Based on both studies it is clear that recovery of riparian areas using seedlings of pioneer species inoculated with mycorrhizal fungi represents an important strategy to increase plant biomass accumulation and, consequently, carbon sequestration.

In the last decade, works have evaluated the relation of arbuscular mycorrhizal colonization on plant biomass production and nutrient uptake of several tropical woody species. Carneiro et al. (1996) evaluated root colonization and mycorrhizal dependency of 31 species and observed that the latter was affected by phosphorus addition. Nevertheless, inoculation with AMF and addition of superphosphate acted synergistically on initial growth of 90% of the species studied. Using an assemblage of AMF as inoculum, Pouyú-Rojas and Siqueira (2000) observed that inoculation of seven woody species at nursery or at transplanting increased growth response and nutrition after transplant into the field and Carneiro et al. (2004) found that inoculated *Cecropia pachystachya* had higher survival and growth after transplanting compared to non-mycorrhizal seedlings. Conversely, Vandresen et al. (2007) observed that AMF inoculation and fertilization had no effect on seedling survival after transplanting, a result that could be partially explained by using an assemblage of fungal species multiplied under trap cultures conditions and stored for two years in refrigerator before inoculation. Compatibility studies of AMF isolates and tropical woody species has demonstrated that isolates of *Glomus clarum, Entrophospora colombiana, Scutellospora pellucida* and *Glomus etunicatum* benefited 80% of the 16 plant species studied by Pouyú-Rojas et al. (2006). Relationship between root morphology, soil fertility and sucessional stages with mycorrhizal colonization and mycotrophic growth has been extensively studied by Zangaro and coworkers (Zangaro et al. 2003, 2005, 2007).

In Santa Catarina state, studies with AMF in riparian forests have emphasized their role in improving plant biomass accumulation and carbon sequestration considering different woody species and soil classes. In the Itajai river watershed, permanent flooded areas are occupied by *Organossolos* and *Gleissolos* while higher, well or moderated drained terrains are dominated by *Neossolos* and *Cambissolos*. At greenhouse conditions, Pasqualini et al. (2007) studied plant growth response of pioneer (*Schinus terebinthifolius, Alchornea glandulosa, Cedrella fissilis, Cytharexyllum myrianthum*) and late secondary (*Annona cacans, Cabralea canjerana, Marlierea tomentosa, Magnolia ovata*) woody species to mycorrhizal inoculation and phosphorus addition. Plants were inoculated with assemblage of AMF originating from a *Cambissolo* and *Neossolo*, which correspond to Inceptisols and Entisols according to the US Soil Survey, respectively. The authors categorized mycorrhizal dependency of pioneer species in highly responsive and very highly responsive while late secondary species were independent, marginally or moderately responsive, regardless of soil classes that plant were developed. They also observed that mycorrhizal dependency was inversely correlated with plant seed mass: small-seeded species tended to be more dependent on mycorrhizal than large-seeded species, a pattern also detected by Siqueira et al. (1998) and Zangaro et al. (2000).

Quantities of carbon sequestration by woody plants can be estimated using different approaches: allocation of biomass to distinct root organs (stems, roots, and leaves), physiological measurements of C-rich compounds like carbohydrates and starch, and estimates based on equations considering height and stem diameter. Under greenhouse conditions, AMF inoculation altered the allocation of C compounds measured as soluble carbohydrates according to the host and soil (Figure 1). Pioneer species *Schinus terebinthifolius* Raddi and *Cytharexyllum myrianthum* Cham. were grown in a *Cambissolo* or *Organossolo* and submitted to the following treatments: sterilized soil to obtain a free-mycorrhizal treatment (Control), non-sterilized soils (Native), and non-sterilized soil added with an assemblage of AMF (Inoculated). Plants under Native or Mycorrhizal treatments tended to have larger amounts of shoot soluble carbohydrates compared to Control plants. Inoculation with an assemblage of AMF resulted in large amounts of soluble carbohydrates to the roots in *C. myrianthum* and *S. terebinthifolius* grown in *Cambissolo.* The main conclusion that inoculation significantly changes carbon allocated to shoots and roots is noticeable. More important is that fungal development in Inoculated plants was sufficient to function as carbon sink to roots and certainly provides a pathway for this carbon enter the soil. *Schinus terebinthifolius* is a pioneer species highly recommended to initiate the revegetation process during recovery of riparian forests. We established in greenhouse for 5 months *S. terebinthifolius* seedlings in a non-sterilized *Cambissolo* and inoculated with isolates of *Glomus clarum, Scutellospora heterogama,* and *Acaulospora koskei.* Plants were transplanted to a riparian area under recuperation and after one year at field conditions, plants inoculated with *G. clarum* and *A. koskei* averaged 0.88 Kg C tree^{-1} and 0.87 Kg C tree^{-1}, respectively, compared to plants grown on non-sterile *Cambissolo* that averaged 0.55 Kg C tree^{-1} (Figure 2). This data shows that inoculation of seedlings with selected fungal isolates can influence plant growth and carbon fixation on plant biomass on a long-term basis and therefore, inoculation should be considered a key process on the seedling establishment for revegetation of riparian areas.

a)

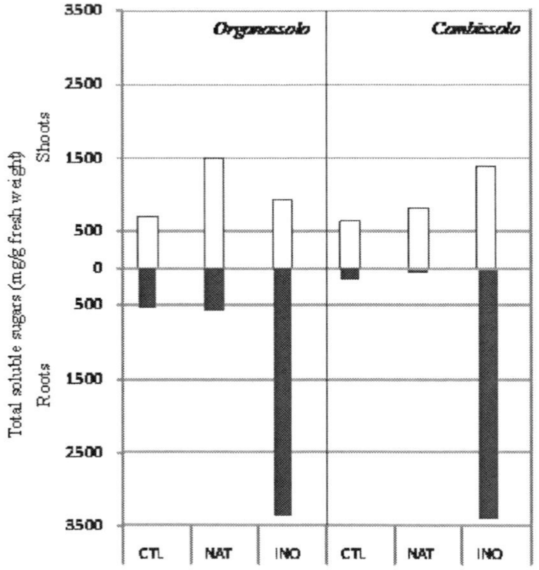

b)

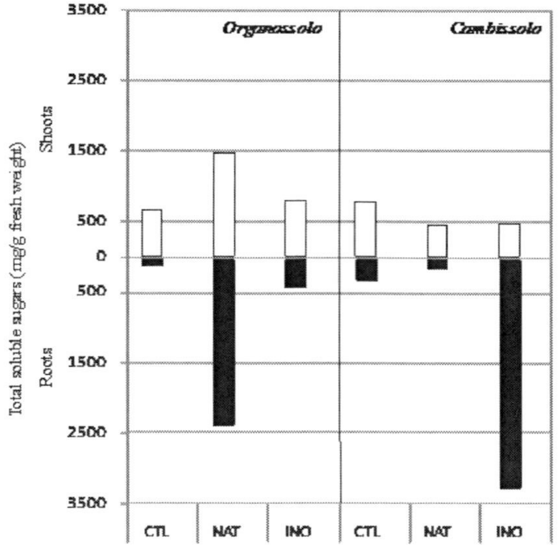

Figure 1. Total soluble sugars (mg/g fresh weight) in (a) Cytharexyllum mirianthum and (b) Schinus terebinthifolius grown in Organossolo and Cambissolo and left non-inoculated (CTL), inoculated with native AMF (NAT), and inoculated with an assemblage of AMF isolates (INO).

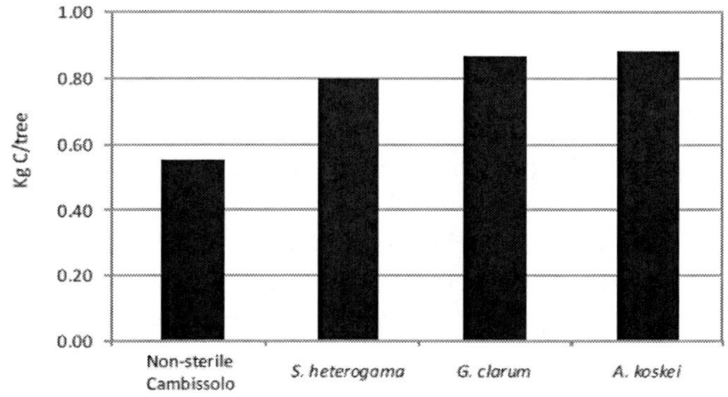

Figure 2. Carbon (Kg C/tree) of Schinus terebinthifolius after one year at field conditions and inoculated with native fungi (non-sterile Cambissolo), Scutellospora heterogama, Acaulospora koskei, and Glomus clarum.

AMF Species Richness in Riparian Environment

Several edaphic factors are known to influence AMF species distribution and abundance including pH (Siqueira et al. 1989, Porter et al. 1987), nutrients (Johnson et al. 1991) and moisture (Anderson et al. 1984, Miller 2000).

In riparian ecosystems, environmental heterogeneity is created by a moisture gradient occurring from the active river channel to inland that is directly related with some factors like the river width, alluvial deposition, and frequency and duration of flooding (Durigan et al. 2000). This gradient impact upon plant community composition and therefore, has the potential to influence mycorrhizal fungal communities in riparian soils. There are few studies in the literature on AMF or ectomycorrhizal fungi in riparian environment and we provide a summary of the main results of these studies (Table 1) and the list of AMF species found in riparian sites (Table 2).

In the United States, AMF occurrence and species diversity in riparian habitats have been studied in the Arizona state. Kennedy et al. (2002) observed the seasonal dynamics of AMF associated with *Sporobulus wrightii* in four riparian sites. They detected fifteen AMF species associated with the host and species richness ranging from 9 to 13 according to the site. Most species detected were *Glomus* followed by *Acaulospora*. Beauchamp et al. (2006) studied the occurrence of AMF species and determined environmental variables related to AMF species richness associated with *Populus* and *Salix* stands.

They found 27 AMF species in trap cultures and fourteen of them pertained to *Glomus* and eight to *Acaulospora*. Other genera detected were *Entrophospora, Paraglomus* and *Archaeospora*. In this study, AMF species richness declined with plant species stand age and distance from the river channel. Interesting is that a subset of all species found tended to occur with high frequency in hydric areas.

Parádi and Baar (2006) in the Netherlands studied the ectomycorrhizal fungal community associated with three homogenous *Salix alba* riparian forest that experienced flooding for up to 144 days during the last ten years. A total of 12 types of ectomycorrhizal fungi were detected using molecular methods; *Tuber* spp. were the most abundant representative in all sites and species of *Hebeloma* and *Cenococcum* were also detected.

Table 1. Published works of arbuscular mycorrhizal fungi and ectomycorrhizae in riparian systems

Reference	*Country/ state*	*Main results*
Carrenho et al. (2001)	Brazil – São Paulo	- AMF species richness was higher under the rhizosphere of *Croton urucurana* and *Inga striata* than *Genipa americana*. - AMF spore number tended to increase with sucessional stages of the plant host. - Mycorrhiza community was dominated by species of *Glomus* and *Acaulospora.* - Diversity indices, richness and evenness for AMF community tended to decrease in the rizosphere of climax species compared to pioneer species. - Most frequent species: *Glomus macrocarpum, G. etunicatum, G. claroideum, Entrophospoa kentinensis,* and *Acaulospora tuberculata*
Kennedy et al. (2002)	USA - Arizona	- Presence of 15 AMF species associated with *Sporobulus wrightii* in four riparian sites. - *Glomus* was the dominant genus associated with *S. wrightii*. - Root colonization was significantly different in plants of *S. wrightii* occupying upper terraces and lower floodplains, but changes in total colonization were not correlated with percentage relative moisture. - Most frequent species: *Glomus mosseae, G. intraradices, G. spurcum,* and *Paraglomus occultum*
Beauchamp et al. (2006)	USA – Arizona	- AMF species richness declined with stand age and distance from and elevation above the channel - The following factors played a role in structuring AMF community: distance from and elevation above the active

		channel, stand age, annual species cover, perennial species richness, and exchangeable potassium concentration - Despite that most AMF species were found across a wide range of soil conditions, a subset of AMF species tended to be more often found in hydric areas. - Most frequent species: *G. intraradices*, *G. microaggregatum, G. spurcum*, *G. eburneum* and *G. mosseae*).
Parádi and Baar (2006)	The Netherlands	- Twelve distinct types of ectomycorrhizal fungi were detected in roots of *Salix alba* and ectomycorrhizal fungal community was dominated by *Tuber* sp. - Basidiomycete fungi pertaining to *Hebeloma* sp. and telephoroid spp. were found mainly in 20-year-old forest with lowest soil nutrient content. - Only a limited number of ectomycorrhizal fungi can resist environmental conditions caused by flooding and drought.

Table 2. Occurrence of AMF fungal species in riparian soils

AM Fungal Species	Carrenho et al. (1997)	Carrenho et al. (2001)	Kennedy et al. (2002)	Beauchamp et al. (2006)	Stürmer et al. (unpublished)
Acaulosporaceae					
Acaulospora capsicula Blaszkowski					X
A. delicata Walker, Pfeiffer and Bloss			X	X	X
A. foveata Trappe and Janos					X
A. koskei Blaszkowski					X
A. longula Spain and Schenck	X	X			
A. mellea Spain and Schenck		X			X
A. morrowiae Spain and Schenck	X		X	X	X
A. scrobiculata Trappe		X	X		X
A. spinosa Walker and Trappe		X		X	
A. tuberculata Janos and Trappe	X	X			

A. undulata Sieverding					X
Entrophospora colombiana Spain and Schenck					X
E. kentinensis Wu and Liu		X			
Glomeraceae					
Glomus claroideum Schenck and Smith		X		X	
G. clarum Nicol. and Schenck		X			X
G. deserticola Trappe, Bloss and Menge				X	
G. eburneum Kennedy, Stutz and Morton			X	X	
G. etunicatum Becker and Gerd.	X	X		X	
G. fasciculatum (Thaxter) Gerd. and Trappe emend. Walker and Koske				X	
G. globiferum Koske and Walker		X			
G. intraradices Schenck and Smith			X	X	
G. invermaium Hall		X			
G. luteum Kennedy, Stutz and Morton			X	X	
G. macrocarpum Tul and Tul.	X	X	X	X	

Table 2. (Continued)

AM Fungal Species	Carrenho et al. (1997)	Carrenho et al. (2001)	Kennedy et al. (2002)	Beauchamp et al. (2006)	Stürmer et al. (unpublished)
G. microaggregatum Koske, Gemma and Oleixa		X	X	X	
G. microcarpum Gerd. and Trappe		X			
G. mosseae (Nicol. and Gerd.) Gerd. and Trappe			X	X	
G. spurcum Pfeiffer, Walker and Bloss emend. Kennedy, Stutz and Morton			X	X	
G. tortuosum Schenck and Smith				X	

Gigasporaceae					
Gigaspora decipiens Hall and Abbott	X				X
G. margarita Becker and Hall		X			
G. ramisporophora Spain, Sieverding and Schenck	X				
Scutellospora calospora (Nicol and Gerd.) Walker and Sanders		X			
S. erythropa (Koske and Walker) Walker and Sanders		X			
S. fulgida Koske and Walker		X			
S. heterogama Nicol. and Gerd.		X			
S. persica (Koske and Walker) Walker and Sanders	X				
Paraglomeraceae					
Paraglomus brasilianum (Spain and Miranda) Morton and Redecker					X
P. occultum Walker		X	X	X	
P. laccatum Renker, Błaszk. and Buscot					X
Archaeosporaceae					
Archaeospora trappei (Ames and Linderman) Morton and Redecker			X	X	X
Ambisporaceae					
Ambispora gerdemanni (Rose, Daniels and Trappe) Walker, Vestberg and Schuessler		X			
Identified species	8	1	13	17	14
Non-identified species	0	21	2	10	13
Total Species Richness	8	22	15	27	27

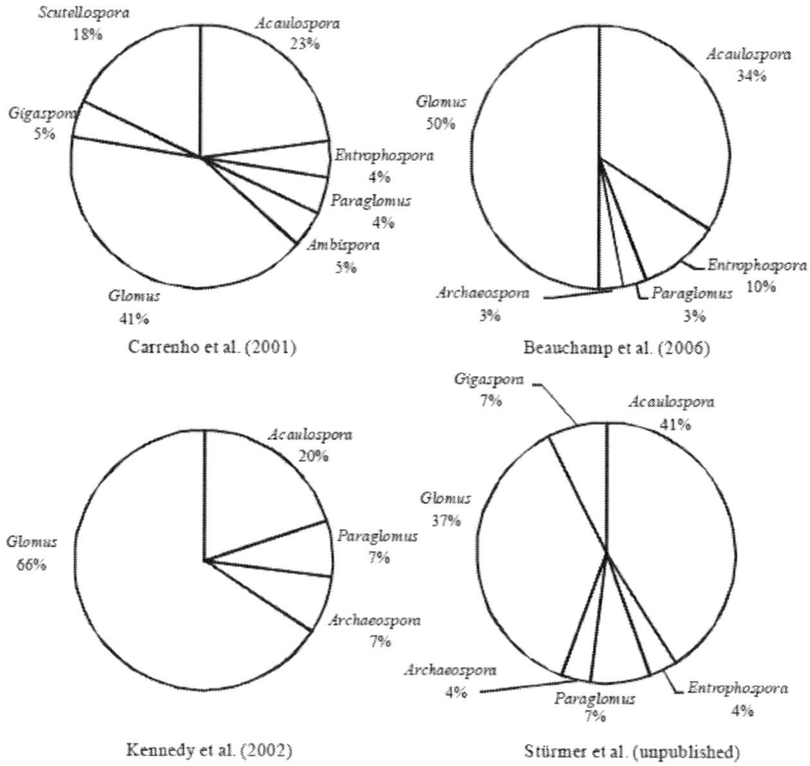

Figure 3. Proportion of genera of glomeromycetes occurring in different studies in riparian soils.

In Brazil, Carrenho et al. (1997) and Carrenho et al. (2001) surveyed the occurrence of AMF species associated with rhizosphere of *Croton urucurana, Inga striata* and *Genipa americana* used in a revegetation process to recuperate riparian areas. In the work of 1997, they found 8 AMF species associated with the host surveyed and no differences on AMF species richness was detected between young and mature plants. Carrenho et al. (2001) found 22 species belonging to *Acaulospora, Glomus, Entrophospora, Gigaspora, Scutellospora,* and *Paraglomus*. Overall *Glomus macrocarpum, G. etunicatum, G. claroideum, E. kentinensis,* and *A. tuberculata* were the most often species recovered. Spore abundance of AMF was correlated with the successional stage: spore numbers were low in the pioneer *C. urucurana* and tended to increase in the climax *G. americana.* In riparian soils in the Itajai river valley we have investigated AMF species richness in soil classes occurring in distinct land use systems. Results from surveys in *Organossolos* and *Gleissolos*, both characterized by being hydromorphic soils, under pasture, forest and agricultural systems have recorded 27 species (Table 2). *Paraglomus laccatum* and *Archaeospora trappei* were the most abundant and frequent species recovered, especially in *Gleissolos*. Both soils shared a high percentage of AMF species detected but land use systems under the same type of soil influenced occurrence of some fungal species. For instance, *Paraglomus laccatum* and *Acaulospora morrowiae* were the most abundant species in agriculture while *A. mellea* and *Acaulospora* sp. were the most often recovered in forest, both under *Organossolo*. Some species were detected exclusively under a particular combination of soil and land use system: *Entrophospora colombiana* was detected in both soils only in pasture lands, and *Gigaspora decipiens* was found only in pasture in *Gleissolos*.

Analysis of fungal species richness from studies of Carrenho et al. (1997), Carrenho et al. (2001), Kennedy et al. (2002), Beauchamp et al. (2006), and surveys of riparian ecosystems in the Itajai river valley demonstrate some pattern of AMF communities in riparian soils. Considering the number of species, *Glomus* and *Acaulospora* are dominant genera and make 64% to 83% of the total number of species detected (Figure 3). On the other hand, members of Gigasporaceae are not commonly found: *Scutellospora* species were detected only by Carrenho et al. (1997) and Carrenho et al. (2001) and only one or two *Gigaspora* species were found per study. *Entrophospora* is a common genus present in riparian forest but represented by different species. *Entrophospora kentinensis* was detected by Carrenho et al. (2001), *E. infrequens* and two undescribed *Entrophospora* was found in soils studied by Beauchamp et al. (2006) while *E. colombiana* was recorded in the Itajai valley riparian soils. All species of *Paraglomus* (*P. occultum, P. laccatum,* and *P. brasilianum*) have been detected in riparian soils and *Archaeospora trappei* is also a common species found in these soils.

Concluding Remarks

In this chapter we provided some evidences that inoculate seedlings of pioneer woody species with arbuscular mycorrhizal fungi is an environmental friendly strategy to pursue fast plant growth at field conditions and speed up the recovery process of riparian areas. Models for recovery of riparian areas consider the pivotal role of pioneer plants which are highly responsive to mycorrhizal fungi and therefore inoculation of pioneer seedlings should turn a common practice in nurseries. Besides impacting initial plant growth, AMF function also as an important way to introduce and maintain carbon in the soil improving carbon sequestration in riparian areas through hyphal turnover, glomalin production, increasing soil aggregation and source of chitin in spores and hyphae.

Studies of occurrence of arbuscular mycorrhizal species reveal that riparian soils and the vegetation established herein support a relatively high species richness of glomeromycetes dominated by the genera *Glomus* and *Acaulospora*. Common species in these environments pertain to *Paraglomus* and *Archaeospora*, two genera that are basal from a phylogenetically within the phylum Glomeromycota. Further works in riparian areas should include the role of external mycelium on increasing soil aggregation and the assessment of mycorrhizal inoculum potential of these soils.

REFERENCES

Ab'Saber, AN. O suporte Geoecológico das Florestas Beiradeiras (Ciliares). In: Rodrigues RR, Leitão-Filho HF (editors). *Matas ciliares: conservação e recuperação.* São Paulo: Editora USP- Fapesp; 2000; 15-25.

Anderson RC; Liberta AE; Dickman LA. Interaction of vascular plants and vesicular-arbuscular mycorrhizal fungi across a soil moisture–nutrient gradient. *Oecologia,* 1984, 64, 111–117.

Barbosa, LM. Considerações gerais e modelos de recuperação de formações cilitares. In: Rodrigues RR, Leitão-Filho HF (editors). *Matas ciliares: conservação e recuperação.* São Paulo: Editora USP- Fapesp; 2000; 289-312.

Beauchmap, VB; Stromberg, JC; Stutz, JC. Arbuscular mycorrhizal fungi associated with Populus-Salix stands in a semiarid riparian ecosystem. *New Phytologist*, 2006, 170, 369-380.

Bigarella, JL. *Estrutura e origem das paisagens tropicais e subtropicais.* Florianópolis: UFSC, 2003.

Caldeira, MVW, LF Watzlawick, MV Schumacher, R Balbinot, CR Sanquetta. Carbono orgânico em solos florestais. In: Sanquetta CR, Watzlawich LF, Balbinot R, Ziliotto MAB, Gomes FS (Orgs.). *As florestas e o carbono.* Curitiba:Imprensa Universitária da UFPR; 2002; 191-213.

Carneiro, MAC; Siqueira, JO; Davide, AC; Gomes, LJ; Curi, N; Vale, FR. Fungo micorrízico e superfosfato no crescimento de espécies arbóreas tropicais. *Scientia Forestalis*, 1996, (50), 21-36.

Carneiro, MAC; Siqueira, JO; Davide, AC. Fósforo e inoculação com fungos micorrízicos arbusculares no estabelecimento de mudas de embaúba (Cecropia pachystachya Trec). *Pesquisa Agropecuária Tropical*, 2004, 34, 119-125.

Carrenho, R; Bononi, VLR; Barbosa, LM. Glomales em áreas de recomposição de mata ciliar de Moji-Guaçu, SP, Brasil. *Hoehnea*, 1997, 24, 107-113.

Carrenho, R; Trufem, SFB; Bononi, VLR. Fungos micorrízicos arbusculares em rizosferas de três espécies de fitobiontes instaladas em área de mata ciliar revegetada. *Acta botânica brasílica*, 2001, 15, 115-124.

Desjardins, RL; Smith, W; Grant, B; Campbell, C; Riznek, R. Management strategies to sequester carbon in agricultural soils and to mitigate greenhouse gas emissions. *Climatic Change*, 2005, 70, 283-297.

Durigan, G. Floristica, fitossociologia e produção de folhedo em matas ciliares da região oeste do Estado de São Paulo. *Tese de Doutorado*. Instituto de Biologia - Universidade Estadual de Campinas. Campinas, SP, 1994. 149p.

Durigan, G; Rodrigues, RR; Schiavini, I. A heterogeneidade ambiental definindo a metodologia de amostragem da floresta ciliar. In: Rodrigues RR, Leitão-Filho HF (editors). *Matas ciliares: conservação e recuperação.* São Paulo: Editora USP- Fapesp; 2000; 159-167.

Jakobsen, I; Rosendahl, L. Carbon flow into soil and external hyphae from roots of mycorrhizal cucumber plants. *New Phytologist*, 1990, 115, 77-83.

Johnson, NC; Zak, DR; Tilman, D; Pfleger, FL. Dynamics of vesicular arbuscular mycorrhizae during old field succession. *Oecologia,* 1991, 86, 349-358.

Kageyama, PY; Biella, LC; Palermo Jr., A. Plantações mistas com espécies nativas com fins de proteção a reservatório. In: Anais...Congresso Florestal Brasileiro, 6, Campos do Jordão. Sao Paulo:*Sociedade Brasileira de Silvicultura*, 1990, 1, 109-112.

Kennedy, LJ; Tiller, RL; Stutz, JC. Associations between arbuscular mycorrhizal fungi and Sporobolus wrightii in riparian habitats in arid South-western North America. *Journal of Arid Environments*, 2002, 50, 459-475.

Kozlowski, TT. Physiological-ecological impacts of flooding on riparian forest ecosystems. *Wetlands*, 2002, 22, 550-561.

Lima, WP; Zakia, MJB. Hidrologia de Matas Ciliares. In: Rodrigues RR, Leitão-Filho HF (editors). *Matas ciliares: conservação e recuperação.* São Paulo: Editora USP- Fapesp; 2000; 33-44.

Lobo, PC; Joly, CA. Aspectos ecofisiológicos da vegetação de Mata ciliar do Sudeste do Brasil. In: Rodrigues RR, Leitão-Filho HF (editors). *Matas ciliares: conservação e recuperação.* São Paulo: Editora USP- Fapesp; 2000; 143-157.

Macedo, AC. Revegetação: matas ciliares e de proteção ambiental. São Paulo: Fundação Florestal; 1993.

Marabesi, MA. Efeito do alto CO_2 no crescimento inicial e na fisiologia da fotossíntese em plântulas Senna alata (L.) Roxb.. Dissertação (Mestrado) - Instituto de Botânica da Secretaria de Estado do Meio Ambiente, São Paulo, 2007.

Marcene EA, Dalla Corte AP, Sanquetta, CR, Schneider, CR. Variações nos teores e estoques individuais de carbono fixado com o crescimento de Gmelina arborea Rosb. na região litorânea do Paraná, Brasil. *Revista Scientia Florestalis*, 2006, 71, 55-63.

Miller, SP. Arbuscular mycorrhizal colonization of semi-aquatic grasses along a wide hydrologic gradient. *New Phytologist,* 2000, 145, 145–155.

Naiman, RJ; Décamps, H. The ecology of interfaces: Riparian Zones. *Annual Review of Ecology and Systematics*, 1997, 28, 621–58.

Parádi, I; Baar, J. Mycorrhizal fungal diversity in willow forests of different age along the river Waal, The Netherlands. *Forest Ecology and Management*, 2006, 237, 366-372.

Pasqualini, D; Uhlmann, A; Stürmer, SL. Arbuscular mycorrhizal fungal communities influence growth and phosphorus concentration of woody plants species from the Atlantic rain forest in South Brazil. *Forest Ecology and Management*, 2007, 245, 148-155.

Porter, WM; Robson, AD; Abbott, LK. Factors controlling the distribution of vesicular arbuscular mycorrhizal fungi in relation to soil pH. *Journal of Applied Ecology,* 1987, 24, 663–672.

Pouyú-Rojas, E; Siqueira, JO. Micorriza arbuscular e fertilização do solo no desenvolvimento pós-transplante de mudas de sete espécies florestais. *Pesquisa Agropecuária Brasileira*, 2000, 35, 103-114.

Pouyú-Rojas, E; Siqueira, JO; Santos, JGD. Compatibilidade simbiótica de fungos micorrízicos arbusculares com espécies arbóreas tropicais. *Revista Brasileira de Ciência do Solo*, 2006, 30, 413-424.

Rees, RM; Bingham, IJ; Baddeley, JA; Watson, CA. 2005. The role of plants and land management in sequestring soil carbon in temperate arable and grassland ecosystems. *Geoderma*, 2005, 128, 130-154.

Rillig, MC; Wright, SF; Nichols, KA; Schmidt, WF; Torn, MS. Large contributions of arbuscular mycorrhizal fungi to soil carbon pools in tropical forest soils. *Plant Soil,* 2001, 233,167-177.

Rodrigues, RR. Análise de um Remanescente de vegetação natural às margens do Rio Passa-Cinco, Ipeúna, SP. Tese (Doutorado) - Instituto de Biologia. Campinas, *UNICAMP*. 1992. 373p.

Rodrigues RR; Filho, HF. Matas ciliares: conservação e recuperação. São Paulo:Edusp-Fapesp; 2000.

Siqueira, JO; Colozzi-Filho, A; Oliveira, E; Schenck, NC. Ocorrência de micorrizas vesículo-arbusculares em agro e ecossistemas naturais do Estado de Minas Gerais. *Pesquisa Agropecuária Brasileira*, 1989, 24, 1499-1506.

Siqueira, JO; Carneiro, MAC; Curi, N; Rosado, SCS; Davide, AC. Mycorrhizal colonization and mycotrophic growth of native woody species as related to successional groups in Southeastern Brazil. *Forest Ecology and Management,* 1998, 107, 241–252

Smith, SE; Read, DJ. *Mycorrhizal Symbiosis*. 2nd editions, San Diego: Academic Press; 1997.

Stulen, I; den Hertog, J. 1993. Root growth and functioning under atmospheric CO_2 enrichment. *Vegetatio*, 1993, 104/105, 99-115.

Treseder, KK; Allen, MF. Mycorrhizal fungi have a potential role in soil carbon storage under elevated CO_2 and nitrogen deposition. *New Phytologist,* 2000, 147,189-200.

Van de Geijn, SC; van Veen, JA. Implications of increased carbon dioxide levels for carbon input and turnover in soils. *Vegetatio*, 1993, 104/105, 283-292.

Vandresen, J; Nishidate, FR; Torezan, JMD; Zangaro, W. Inoculação de fungos micorrízicos arbsuculares e adubação na formação e pós-transplante de mudas de cinco espécies arbóreas nativas do sul do Brasil. *Acta botânica brasílica*, 2007, 21, 753-765.

Wilson, GWT; Rice, CW; Rillig, MC; Springer, A; Hartnett, DC. Soil aggregation and carbon sequestration are tightly correlated with the abundance of arbuscular mycorrhizal fungi: results from long-term field experiments. *Ecology Letters*, 2009, 12,452-461.

Zangaro, W; Bononi, VLR; Trufem, SB. Mycorrhizal dependency, inoculum potential and habitat preference of native woody species in South Brazil. *Journal of Tropical Ecology*, 2000, 16, 603-622.

Zangaro, W; Nisizaki, SMA; Domingos, JCB; Nakano, EM. Mycorrhizal response and successional status in 80 woody species from south Brazil. *Journal of Tropical Ecology*, 2003, 19:315-324.

Zangaro, W.; Nishidate, FR; Camargo, FRS; Romagnoli, GG; Vandresen, J. Relationships among arbuscular mycorrhizas, root morphology and seedling growth of tropical native woody species in southern Brazil. *Journal of Tropical Ecology*, 2005, 21, 529-540.

Zangaro, W; Nishidate, FR; Vandresen, J; Andrade, G; Nogueira, MA. Root mycorrhizal colonization and plant responsiveness are related to root plasticity, soil fertility and successional status of native woody species in southern Brazil. *Journal of Tropical Ecology*, 2007, 23, 53-62.

Zhu, Y-G; Miller, RM. Carbon cycling by arbuscular mycorrhizal fungi in soil-plant systems. *TRENDS in Plant Science*, 2003, 8,407-409.

Wang, B; Qiu Y-L. Phylogenetic distribution and evolution of mycorrhizas in land plants. *Mycorrhiza,* 2006, 16, 299-363.

WWF. 2003. A água, as cidades e as florestas. Disponível em: *http://www.wwf.org.br.*

INDEX

A

B

C

D

E

F

M

N

O

P

Q

R

S